CAMBRIDGE MONOGRAPHS IN
EXPERIMENTAL BIOLOGY
No. 19

EDITORS:
P. W. BRIAN
J. W. S. PRINGLE

DEFENCE MECHANISMS OF PLANTS

THE SERIES

DEFENCE MECHANISMS
OF
PLANTS

BY

BRIAN J. DEVERALL

Professor of Plant Pathology, University of Sydney

CAMBRIDGE UNIVERSITY PRESS

CAMBRIDGE

LONDON · NEW YORK · MELBOURNE

Published by the Syndics of the Cambridge University Press
The Pitt Building, Trumpington Street, Cambridge CB2 1RP
Bentley House, 200 Euston Road, London NW1 2DB
32 East 57th Street, New York, NY 10022, USA
296 Beaconsfield Parade, Middle Park, Melbourne 3206, Australia

© Cambridge University Press 1977

First published 1977

Printed in Great Britain
at the
University Printing House, Cambridge

Library of Congress Cataloguing in Publication Data
Deverall, Brian J.
Defence mechanisms of plants.
(Cambridge monographs in experimental biology; no. 19)
Bibliography: p.
Includes index.
1. Plants – Disease and pest resistance. 2. Host–
parasite relationships. I. Title. II. Series.
SB750.D48   581.2'32   76–12917
ISBN 0 521 21335 5

Contents

Preface

This book is concerned with the dynamic mechanisms involved in the defence of plant cells against attack by parasitic bacteria and fungi. Thus I scarcely discuss those plant features such as bark and cuticle which play an obvious role in defence, but which are essentially static contributors. Circumvent these barriers and the ability of apparently undifferentiated parenchyma to defend itself is revealed. Furthermore, this ability is dependent upon particular genes in plant and parasite which interact after infection. My interest is with the processes by which plant cells perceive the approach of an intruder and occasionally permit, but commonly discourage, its further progress. How do the genes of host and parasite communicate to determine the outcome of attempted parasitism? Is there a universal defence mechanism in all plants, and, if so, what is it? What contribution does the much studied process of phytoalexin formation make to the defence of plants?

Research on the physiology of host–parasite relationships has been prolific in recent years and a number of multi-author treatises are being published on different aspects of this work. Hopefully, this monograph will make a useful contribution by presenting a shorter and personal view of those parts of this research which bear directly upon the processes of resistance in plants. My envisaged readership comprises research workers in the subject, and University teachers and their advanced students in plant pathology, botany and plant biochemistry.

I wish to thank Professor A. H. Ellingboe for his suggestions and comments on some of the chapters, Elizabeth Froggatt for typing the manuscript, Stella McLeod for assistance with the References and my family for their tolerance and encouragement.

University of Sydney　　　　　　　　　　　　　B. J. DEVERALL
February 1976

CHAPTER I

Introduction to the Host–Parasite Interaction

Plants at all stages of their life-cycles are exposed to many potentially parasitic micro-organisms. Seeds germinate in soils which contain numerous resting parasites awaiting the arrival of roots to stimulate them into activity. Aerial parts of plants are inoculated by fungal spores and bacterial cells carried in air currents and rain-splash droplets. Under favourable conditions of moisture and temperature, plant tissues are thus subjected to attempted infection on numerous occasions. However, these attempts often fail, and most plants remain healthy. Successful establishment of a parasite depends upon a special genetical and physiological relationship so that the cells of the host accept the parasite.

This book is concerned with the processes whereby plants succeed in remaining healthy despite their constant exposure to potential parasites. It leads to a consideration of one of the most interesting problems in biology and biochemistry, namely the molecular basis of the high degree of specialization which is often observed in relationships between parasites and hosts. As will become apparent, there are reasons to believe that the basis of much specialized parasitism rests in the ability of parasites to confound a recognition system linked to other reactions in host cells. Through this linked system, the host normally notices and then fails to accept the intrusion of an alien organism.

INFECTION TYPE, SUSCEPTIBILITY AND VIRULENCE

Before starting to analyse these processes, it is essential to re-emphasize that a disease is the product of the interaction of two organisms, host and parasite. This product can depend upon fine differences in the properties of the two organisms, as is well recognized in research on rust diseases. Thus under

TABLE 1. *The alternative attributes of parasite and host and the result of their interaction*

Attributes of parasites	Attributes of host plants	Resulting infection type
Virulence	Susceptibility	High
Lower virulence	Lower resistance	Intermediate
Avirulence	Resistance	Low

standard and ideal conditions for infection, an isolate of a wheat rust fungus will fail to develop on one cultivar of wheat *Triticum aestivum*, but will produce large uredia on another and an intermediate condition on a third cultivar, comprising, for example, small uredia surrounded by an area of necrosis or chlorosis in the host. These different products are termed *infection types*, and they are determined by the genetic constitutions of the rust isolate and wheat cultivar under particular environmental conditions. A common convention is to qualify infection type by *high* where there is substantial rust development and by *low* where there is little or no development (Loegering, 1966). A low infection type implies that the cultivar was *resistant* to the rust isolate and that the isolate was *avirulent* on that cultivar. High infection type implies the product of an interaction between a *susceptible* host and a *virulent* rust isolate. Thus, resistance or susceptibility as properties of a plant are defined with respect to the response of that plant to infection by a particular isolate of parasite. Similarly, virulence or avirulence as properties of a parasite are defined with respect to success or failure to colonize a particular host plant. Intermediate infection types result from interactions between parasites and hosts possessing lower degrees of virulence and resistance. The interaction between these alternative attributes of parasites and hosts is shown in Table 1.

The concept of infection type will be used in considering all host–parasite interactions in this book, and will not be restricted to the rust diseases.

Many virulent parasites such as most rust fungi grow in an apparently harmonious relationship with their susceptible hosts because they cause no visible adverse reaction in the host cells which they penetrate. It is generally assumed that these parasites derive their nutrients from the living host cells, and they are thus said to be *biotrophic* in their parasitism. By contrast, soft-rot parasites such as the bacterium *Erwinia* in tubers of the potato *Solanum tuberosum* or leaf-spot parasites such as *Botrytis fabae* in the leaves of the broad bean *Vicia faba* kill adjacent plant cells by chemical secretions. These highly successful parasites presumably derive their nutrients from cells which they have killed and thus are said to be *necrotrophic* in their parasitism.

The rust fungi and *B. fabae* are extremely different types of parasite in their relationships with host cells, but it is important to appreciate that some parasites, such as *Colletotrichum lindemuthianum* in bean *Phaseolus vulgaris*, are often biotrophic for part of their development and necrotrophic for the remaining part. Analyses of the features essential to the host–parasite interaction must consider these possibly changing relationships. Very special capacities of the parasite might be anticipated during its biotrophic phase when the integrity and function of living host cells are maintained despite the intrusion of hyphae or haustoria into protoplasts. During this phase, the relationships between the cells of host and parasite can be termed compatible. *Compatibility* will be used in this book to describe harmonious relationships between parasite and host, and *incompatibility* to describe interactions which cause deleterious changes in cells of host and/or parasite.

The use of the concept of compatibililty is thus more restricted here compared with its use by Day (1974) to refer to any host–parasite interaction which gives rise to a high infection type. This restriction might facilitate an analysis of the nature of the compatibility which enables cells of a biotroph and host to live together harmoniously. Quite a different sequence of physiological events might be envisaged to underlie the success, as a parasite, of the virulent necrotroph *B. fabae* in giving rise to a high infection type despite the incompatibility of broad bean cells to the fungus. The concept of compatibility in the host–parasite cellular relationships should therefore be kept distinct from the concept of final infection type.

No attempt is made here to give a detailed review of the genetics of the host–parasite interaction, and the reader is referred to the book bearing that title by Day (1974) and to reviews by Ellingboe (1976), Johnson (1976*a, b*) and Day (1976) for more complete discussions of this subject. However, as part of a general introduction to the nature of the host–parasite interaction, it is well to recall the widely held generalizations pertaining to the genetic control of the interaction. These arise from results of analyses done by Flor on the inheritance of resistance and virulence in flax rust disease. Different genes conferred resistance in cultivars of flax *Linum usitatissimum* to different races of the rust, and different genes conferred avirulence in rust races to different flax cultivars. The analyses were the basis for the gene-for-gene hypothesis of Flor (1956) which implies that infection type is determined by complementary single genes for resistance in the host and avirulence in the parasite.

TABLE 2. *Dependence of infection type upon genes in host and parasite*

		Parasite alleles for avirulence	
		A	*a*
Host alleles for resistance	*R*	Low	High
	r	High	High

The gene-for-gene hypothesis is considered to apply to many host–parasite interactions, and its basis is illustrated in the simplest form in Table 2 where infection type depends upon the alleles at single gene loci in parasite and host. A low infection type results from the genetic interaction between the dominant alleles for avirulence in the parasite and for resistance in the host. Absence of the dominant allele in either partner results in a high infection type where virulence and susceptibility are expressed. Thus expression of avirulence and resistance is a more particular phenomenon, requiring precise matching of genetic information in both partners.

TABLE 3. *Dependence of infection type upon complementary genes in host and parasite*

		Alleles at gene loci for avirulence in the parasite			
		A1 A2	A1 a2	a1 A2	a1 a2
Alleles at	R1 R2	Low	Low	Low	High
gene loci	r1 R2	Low	High	Low	High
for resistance	R1 r2	Low	Low	High	High
in the host	r1 r2	High	High	High	High

The complementary role of alleles at specific gene loci in control of infection type is illustrated in Table 3, where two gene loci are depicted in both host and parasite. Here it can be seen that low infection type results from the matching of specific complementary dominant alleles in host and parasite. Thus avirulence and resistance are expressed only when avirulence gene A1 matches resistance gene R1 or when A2 matches R2. Virulence and susceptibility are expressed if avirulence gene A1 is matched with recessive allele r1 or dominant allele R2 in the host. Thus Table 3 emphasizes more strongly than Table 2 that expression of avirulence and resistance is the result of a highly specific genetic interaction.

Many analyses show that resistance in crop plants is inherited as single dominant genes, but relatively few corresponding analyses of inheritance of virulence in parasites have been performed. Putative genes for avirulence are often assigned to races of parasites based on the gene-for-gene hypothesis and knowledge of corresponding genes for resistance in the host. However, exceptions to the major generalizations discussed above are known where susceptibility and virulence are the dominant characters in host and parasite; for example, the relationship between oats *Avena sativa* and the fungus *Helminthosporium victoriae* which will be considered in a later chapter. Polygenic control of resistance towards particular parasites has been claimed in a number of important crop plants, but this phenomenon requires critical genetic analysis under controlled conditions of environment and, in at least two cases, has not withstood this test (see Ellingboe, 1976; Johnson, 1976*b*).

The genetic control of parasitism in natural vegetation is not well known, and we can only speculate about how resistance and avirulence evolved before man sought resistance genes for incorporation into his crop plants. The selective advantage of resistance to plants seems self-evident, but the role of avirulence in the development of populations of parasites is more difficult to conceive except as a means of separating evolving populations within species.

The most important implications for physiological and biochemical analyses of expression of resistance and avirulence to arise from genetic studies are that infection type is under control of genes in host and parasite, and that the expression of these interdependent properties is often determined by highly specific interactions between particular complementary genes in both partners. Ellingboe (1976) and Johnson (1976*b*) have pointed out that the most specific interactions are usually for expression of resistance and avirulence. The simplest mechanism mediating this expression would be based upon confrontation of primary products of the genes for resistance and avirulence following upon infection. Evidence bearing upon these hypothetical products will be discussed later, but firstly it is useful to consider what is known about stages in parasitism when resistance is expressed. This will be done by assessing the fate of potential parasites at the stages of attempted entry and then early growth into plants. Attempts to understand natural processes of defence in plants should accommodate reports that plants can be cross-protected against normally virulent parasites by previous infection by other organisms. The extent will be sought to which cross-protection is achieved by activation of a process of induced resistance in plants. Any analysis of the process of expression of resistance must assess the contribution of the many anti-microbial chemical compounds in plants, and especially of those which form after infection. Finally, however, it is essential to return to the initial question concerning the nature of recognition between parasites and plants, and to consider the molecular means by which genes in parasite and host interact to determine specificity.

6

CHAPTER 2

Discriminatory Events before and during Penetration into Plants

Many micro-organisms are dispersed in air currents or in splash droplets caused by rain, and thereby arrive on leaves and stems. Other fungi and bacteria move in soil water before encountering roots or persist as resting stages until roots grow into their vicinity. Plants can then influence micro-organisms around their surfaces by physical and chemical means, thus starting an interaction which must be followed by entry of the parasite into the plant by a specialized route, if parasitism is to have a chance of success. This chapter is concerned with the few attempts that have been made to assess quantitatively the contributions to the success or failure of parasitism of these primary interactions between potential parasites and hosts.

EFFECT OF ROOTS ON PARASITES IN THE SOIL

The principal effect of roots on organisms in the soil is a general stimulation of germination and growth, and this is particularly important for parasites most of which remain dormant until contacted by their living substrates. Fungal parasites lie dormant in soil as a number of different types of resting body such as sclerotia, oospores, chlamydospores, basidiospores and hyphal fragments. Their dormancy is considered to be of two types, constitutive or exogenous (Sussman, 1966). Constitutive dormancy is thought to be maintained by internal factors in the fungus, and is particularly important in basidiospores. Experimentally, constitutive dormancy can be broken, in at least a small proportion of spores, by temperature shocks, treatment with certain chemicals, and proximity to other micro-organisms and some plant roots in culture. Presumably environmental fluctuations in soils cause some of these spores to germinate in favourable seasons. It is also likely that emanations from roots induce

7

breaking of dormancy of basidiopsores of mycorrhizal fungi (Fries, 1966). Exogenous dormancy is conferred by external factors in the soil. Among the emanations from other soil inhabitants which prevent germination of fungi are ethylene (Smith, 1973; Smith & Cook, 1974) and possibly antibiotics, many of which were discovered as products of soil fungi and soil Actinomycetes in culture, although their role in natural soils remains uncertain. The effects of some of the fungistatic factors in soil are overcome by stimulatory substances which diffuse from plant roots. Many substances exude from the zone of elongation of roots in particular (Schroth & Hildebrand, 1964) and these include mineral salts, sugars, amino acids, organic acids, nucleotides and vitamins (Rovira, 1965).

Reports of stimulatory emanations from plant roots are plentiful, but toxic factors such as the cyanogenic glucoside linamarin in *Linum* (Trione, 1950), and compounds inhibitory to nematodes in *Asparagus* (Rohde & Jenkins, 1958) and *Tagetes* (Winoto Suatmadji, 1969) are also known to occur in roots. These factors or their anti-microbial products may diffuse in the soil. The role of inhibitors in the pre-penetration interactions between roots and parasitic micro-organisms is not well established, as discussed below, but Wallace (1973) considers that such factors may influence parasitic nematodes before they reach the roots of some plants, although he states that resistance to nematodes usually occurs during or after penetration.

There is as yet little evidence that susceptible roots are exceptionally stimulatory to parasitic fungi in the soil or that resistant roots are repressive. As Schroth & Hildebrand (1964) emphasize, the soil environment is complex, and stimulants and inhibitors diffusing from plant roots are likely to affect the activities of many soil inhabitants, including saprophytic antagonists, with unpredictable consequences for the success or failure of a parasitic organism. Some laboratory experiments, which precluded these complexities, indicated that susceptible roots stimulate the formation of infection structures and that resistant roots suppress the germination of spores in a selective way, but the specificity of these effects has been discounted in subsequent work. Thus Flentje (1957) and Kerr & Flentje (1957) showed that the formation of appressoria of *Pellicularia filamentosa* was stimulated by exudates from susceptible roots, but Flentje, Dodman & Kerr (1963) and de Silva & Wood (1964) concluded that there

8

was little difference between hosts and non-hosts in this phenomenon. Buxton (1957) found that washings from resistant cultivars of the pea *Pisum sativum* inhibited the germination of conidia of *Fusarium oxysporum* f. sp. *pisi*, and then revealed a similar phenomenon with washings from resistant cultivars of the banana *Musa* sp. and conidia of *F. oxysporum* f. sp. *cubense* (Buxton, 1962). However, specificity in the effects of exudates from pea roots on the germination of micro-conidia or chlamydospores of *F. oxysporum* f. sp. *pisi* was not found in the work of Kommedahl (1966). Furthermore, sterile exudates from resistant and susceptible cultivars were equally stimulatory to different physiologic races of the parasite *in vitro* and also when allowed to diffuse through porous blocks into soil containing chlamydospores (Whalley & Taylor, 1973). There is therefore little reason to believe that susceptible roots selectively encourage germination and chemotropic growth of virulent parasites.

The attraction of zoospores of several *Phytophthora* spp., *Aphanomyces euteiches*, *Olpidium* spp. and *Pythium aphanidermatum* to the zones of elongation of different roots has been established, as discussed by Hickman & Ho (1966). Zoospores swim towards, attach to and encyst on the roots. Water-soluble extracts of roots were attractive to zoospores of *P. aphanidermatum* (Royle & Hickman, 1964*a*, *b*) as also were many of the individual compounds known to exude from roots. The only substance in this work which caused chemotaxis, trapping and encystment of zoospores was glutamic acid in the presence of ammonium bases. However, glutamic acid was considered unlikely to act alone in attracting zoospores to roots. Troutman & Wills (1964) found that zoospores of *Phytophthora parasitica* were attracted to negative electrodes, and suggested that electrostatic adhesion might occur to the zone of elongation of roots. Contrary to most research on the response of zoospores to roots, Zentmyer (1961) saw that some zoospores were specifically attracted to host roots and not to non-host roots. *P. cinnamomi* moved towards roots of susceptible cultivars of the avocado *Persea americana* within a few minutes, and encysted and germinated within an hour. Zoospores of this fungus moved less readily to roots of resistant cultivars and were not attracted to roots of the tomato *Lycopersicon esculentum*, tobacco *Nicotiana tabacum* and *Citrus* sp. Similarly, zoospores of *P. citrophthora* were attracted to roots of citrus but not to those of avocado. In further work, Khew & Zentmyer (1973) found that zoospores of five species of *Phytophthora* including *P. cinnamomi* and *P. citrophthora* responded in a similar chemotactic way to many compounds known to exude from plant roots. Later Khew & Zentmyer (1974) showed that zoospores of seven species including *P. parasitica* responded electrotactically to positive, and not negative, electrodes, contrary to the report by Troutman & Wills (1964). With the awareness of so many non-specific attractants to roots, it is difficult to understand the nature of any specific taxis to susceptible roots.

Most evidence, therefore, suggests the non-specific stimulation of activity in soil micro-organisms by plant roots, but there are a few reports, requiring substantiation, that some events at this pre-penetration stage might determine resistance or susceptibility of the host plant to soil-borne plant parasites. Verification of reports of specific interactions should desirably be based upon the more difficult experimentation in the complex environment of natural soils, and not solely on the necessary testing of phenomena in model situations in the laboratory.

INTERACTIONS WITH PARASITES ON AERIAL PARTS OF PLANTS

The early investigations of Brown (1922) suggested that leaves and petals act as sources of stimulants and inhibitors of fungal development in overlying infection droplets. Water droplets were shown not only to change in electrical conductivity during incubation on plant parts but also to accumulate compounds which affected spore germination of *Botrytis cinerea*. These observations suggest that aerial parts of plants might influence the behaviour of micro-organisms on surfaces in much the same way as roots affect microbial activity. However, recent research has revealed that some of these effects might be caused by other organisms on the plant surface.

Bacteria multiply rapidly in droplets of water on leaf surfaces, especially in the presence of fungal spores, and some of these bacteria produce inhibitors of germination of fungi on the leaf (Blakeman & Fraser, 1971). Irradiation of water droplets on leaves of beetroot *Beta vulgaris* and *Chrysanthemum* sp. with high-intensity ultra-violet light failed to decrease bacterial populations (Sztejnberg & Blakeman, 1973). Because similar bacteria were eliminated in control droplets on glass slides, this suggests that bacteria are sheltered from irradiation on the leaf, either persisting in intercellular spaces or beneath folds on the cuticle. Bacteria may therefore be endemic to leaf surfaces, and can clearly multiply and influence potential parasites of plants.

Pollen grains can act as major sources of stimulants to the germination of spores of many fungi, particularly necrotrophic parasites such as *Botrytis* and *Fusarium* spp. The remarkable effect of pollen was shown following experiments on infection of fruit of the strawberry *Fragaria ananassa* by *B. cinerea* (Chou & Preece, 1968). When anthers were removed before spraying the fruit with conidia, no infection occurred, but in the presence of anthers rapid rotting of the fruit took place. Pollen and pollen extracts were then shown to promote germination of conidia and infection of strawberry petals and broad bean leaves. Thus emanations from pollen confer virulence on *B. cinerea*, normally avirulent on all but ageing flowers and leaves. Wheat anthers stimulate the infection of wheat spikelets by *F. graminearum* (Strange & Smith, 1971). The fact that spikelets are resistant before anthesis, or if the anthers are removed, suggests that the substances which diffuse from the anthers to stimulate the fungus are essential contributors to the virulence of the fungus on this part of the wheat plant. Choline and betaine have been identified as two major components of wheat anthers that stimulate germination of *F. graminearum* (Strange, Majer & Smith, 1974), but it is not yet confirmed that they are the virulence factors in anthers. Anthers and fallen pollen grains may be contributory factors to the development of outbreaks of certain types of disease under wet conditions when inoculum is present. Extensive experimentation on changes in the myco-flora of the leaves of rye *Secale cereale*, as a result of exposure to pollen under natural and artificial conditions, shows that competitive and antagonistic fungi may be stimulated also (Fokkema, 1968, 1971). As a consequence, parasitism is not necessarily promoted by anthesis under field conditions, and the extent of the significance of anthesis in disease development remains for wider investigation.

Following a number of demonstrations of interactions between parasitic and saprophytic fungi and bacteria on the leaf surface, the ecology of the leaf surface is now recognized as an important subject for research (Preece & Dickinson, 1971).

11   2-2

TABLE 4. *Percentage of spores of avirulent and virulent* Colletotrichum lindemuthianum *giving rise to appressoria on resistant and susceptible bean hypocotyls (from Skipp & Deverall, 1972)*

Hours from inoculation	Low infection type (%)	High infection type (%)
24	21	12
48	45	45
72	53	43

These ecological interactions can also be influenced by the underlying cuticle, which is a complex material with many components and an intricate surface configuration. Some cuticles may permit movement of molecules from beneath, thus affecting parasites on the surface. Other cuticles yield antifungal substances upon extraction with organic solvents, and it is possible that these substances may interfere with attempted parasitism either by diffusion into infection droplets or during penetration. However, Martin & Junifer (1970) in their comprehensive account of plant cuticles were reluctant to conclude that cuticles had much direct inhibitory or stimulatory effect on potential parasites. They considered that resistance might be conferred indirectly by physical effects on infection droplets. The chemical composition of the cuticular surface may deter wetting and minimize moisture retention, so affecting the establishment of fungal parasites, most of which require persistence of moisture films for several hours during germination and penetration.

The physical nature of the cuticular surface or chemical emanations from particular sites may influence location of points of entry for parasites. Thus Dickinson (1964) showed that germ-tubes of a rust fungus can detect and grow across ridges on membranes, and suggested that they may do the same on leaf surfaces as they locate stomata. A few attempts have been made to assess the importance of phenomena of these types in resistance. Appressoria of *Colletotrichum lindemuthianum* commonly form on those parts of the cuticle immediately above vertical walls in underlying epidermal cells, but they do this in a similar way on both resistant and susceptible cultivars of bean (Skipp & Deverall, 1972), thus discounting any special role in determining infection type (Table 4).

TABLE 5. *Relationship between zoospore association with stomata and resistance of hop cultivars to* Pseudoperonospora humuli *as measured by yield of sporangia (from Royle & Thomas, 1971)*

Hop cultivars in order of resistance	Zoospores on stomata after 16 hours (%)	Sporangia per leaf disc after 7 days
Ringwood Special	96.8	62600
Eastwell Golding	56.7	60600
1/63/37	76.3	18400
7k 491	78.4	4300
2L 118	91.5	4000

Zoospores of *Pseudoperonospora humuli* are capable of detecting and settling on positive 'Perspex' replicas of stomata on the leaves of the hop *Humulus lupulus*, although relatively slowly. Furthermore, Royle & Thomas (1973) showed that the zoospores had the ability to move very rapidly towards open stomata on the leaf in the light, apparently in response to a chemical stimulus associated with photosynthesis. However, this chemotaxis to stomata occurred similarly on both resistant and susceptible cultivars of hop (Royle & Thomas, 1971), again discounting any special role in regulating virulence and avirulence (Table 5).

Clearly there are many direct and indirect interactions between parasites and aerial parts of plants, but there is not much evidence to support the idea that the fate of avirulent parasites is often determined by inability to establish infections on resistant cultivars. However, relatively little is yet known of the fate in nature of the many parasitic micro-organisms which must alight on a wide variety of leaves and stems under favourable conditions for infection. Many may fail to attempt penetration because they are affected by anti-microbial compounds or are disoriented by the physical contours on the cuticle'. Exudations from pollen can contribute to the susceptibility of plants to some parasites, but again more research is needed to reveal the general significance of this phenomenon to the ecology of the leaf surface and to parasitism.

Entrance into plants via stomata is the most important or only route for many bacteria and most downy mildew and rust fungi. The possibility that failure to gain stomatal entrance might be an important basis of avirulence has been considered by a number of workers, particularly with rusts. Allen (1923) surveyed a number of wheat cultivars resistant to *Puccinia graminis* f. sp. *tritici*, and obtained clear evidence that one cultivar imposed an impediment to entry of infection hyphae via stomatal apertures. However, the same restriction was imposed on races virulent and avirulent on that cultivar. Brown & Shipton (1964) measured the frequency of penetration of stomatal apertures of many wheat cultivars by races of *P. graminis* f. sp. *tritici*. They found remarkable differences in penetration on different cultivars, but these were not related to the final infection type which developed, as shown in Table 6. Perhaps the most remarkable observations were that less than 2 % of the stomata contacted by appressoria were penetrated in some cultivars, yet the rust was able to grow relatively freely once successful entrance was achieved. These results suggest that the combination by plant breeding of other endogenous factors for resistance, with the impediments imposed by some stomata to entrance, could make major contributions to reduction of rust development in wheat.

The evidence of these studies with rust fungi is against the idea that size of rust pustule is controlled by difficulties in entering stomata. Effects of these difficulties on number of rust colonies, as distinct from colony size and host reaction which comprise infection type, do not appear to have been reported.

Assessment of ease of penetration of cuticles by the many fungi which enter by this route is technically difficult, but some pertinent experiments have been done with isolated cuticles and with artificial membranes supplemented with extracts of cuticles. Maheshwari, Allen & Hildebrand (1967) obtained cuticle from *Antirrhinum* leaves and found that it supported normal germination and development of appressoria and infection pegs of several different rusts irrespective of their usual hosts. Yang & Ellingboe (1972) performed a quantitative study of the behaviour of powdery mildew fungi on surfaces of host and non-host cereals, mainly wheat and barley *Hordeum vulgare*. The

TABLE 6. *Relationship between host infection type and percentage of successful penetration from appressoria on wheat varieties inoculated with strain 21-Anz-2 of* Puccinia graminis *f.sp.* tritici (*from Brown & Shipton, 1964*)

Variety	Infection type	Penetrations (%)
Steinwedel	3	19.26
Khapstein 1451	2	17.47
Einkorn 292	;1	17.17
Khapli 12	;1	15.59
Bobbin	3	9.23
Eureka	3	6.63
Emmer II	;1	6.36
Gabo	2–3	4.85
Celebration	3	4.77
Little Club	4	4.48
Mentana	3	4.01
Kanred	0	2.80
Gaza	3	1.85
C.I. 12632	1	1.83
Kenya 117A	2	1.35
II-52-73	2	0.60
Mengavi	1	0.33

fungi produced appressoria freely on the non-host cereals and also on host cereals, even in the presence of specific genes for resistance to the fungal races tested. Similar results were obtained using isolated cuticles, suggesting that there was little involvement of surface contours or chemical composition in resistance or susceptibility. Mere thickness of cuticle does not necessarily present a barrier to infection; for example, the shrub *Euonymus japonicus* has an exceptionally heavy cuticular membrane and yet is highly susceptible to a powdery mildew fungus which penetrates via the cuticle (Roberts, Martin & Peries, 1961). However, the age of cuticles of the poplar *Populus tremuloides* had a marked effect on the ability of the fungus *Colletotrichum gloeosporioides* to penetrate and initiate infections (Marks, Berbee & Riker, 1965), thus showing that some cuticles can present a barrier to penetration. Most investigations based on quantitative comparisons of behaviour on resistant and susceptible plants support the conclusion of Martin (1964) that the contribution of cuticle to resistance is slight compared with the contributions of endogenous interactions after penetration. However, relatively few comprehensive studies have been reported.

15

Penetration of cuticles must be followed by penetration of the epidermal wall by most fungi which enter through intact plant surfaces, except *Venturia* spp. and *Diplocarpon rosae* which grow between the cuticle and the epidermis. Observation of penetration points in epidermal walls, by electron microscopy, such as that of McKeen (1974) studying *Botrytis cinerea*, suggests that this is a neat process accomplished by localized enzyme action by the fungus. Many fungal enzymes are known which can affect different components of walls of plant cells (Wood, 1967). It seems likely that these enzymes are attached to or secreted by tips of infection hyphae during initial penetration of the epidermal wall. An intriguing hypothesis was advanced by Albershiem, Jones & English (1969) to explain avirulence and resistance based on interactions between fungal enzymes and components of epidermal walls. The idea was inspired by (1) their observation that when grown on cell walls as a sole carbon source parasitic fungi produce a sequence of different hydrolytic enzymes and (2) their discoveries that cell walls contain many different polysaccharides. Thus, a sequence of interactions was postulated whereby a wall might be resistant if it released a sugar suppressive to synthesis of a critical enzyme, or if it failed to release a needed sugar. Knowledge of the existence of processes in micro-organisms which can regulate synthesis of enzymes by substrate induction and catabolite repression permits many ingenious explanations of the mechanisms of resistance in the cell wall. The hypothesis requires evidence that avirulent parasites stop growing while attempting to penetrate resistant cell walls. Especially contradictory are the results of an examination by Skipp & Deverall (1972) of the stage in the infection process when avirulent races of *Colletotrichum lindemuthianum* fail in their attempted parasitism of resistant bean cultivars, because this interaction was the model used for development of the hypothesis. Some difficulty was experienced in deciding, by use of light microscopy, whether the fungus stopped before or after penetration of the wall, because necrosis of the underlying cells accompanied expression of resistance and the necrotic cells had granular and brown cytoplasm. However, treatment of the tissue with dilute alkali cleared the browning and granulation from the dead cells and revealed that over 80% of these cells contained short infection hyphae. A parallel study, using electron microscopy, by Mercer, Wood & Greenwood (1974) confirmed that hyphae were present inside the dead protoplasts of some hypersensitive cells. Other comparisons of behaviour of fungi in cells of resistant and susceptible cultivars of several plant species have confirmed that hyphae usually grow equally well through the first cell walls penetrated – for example, the development of *Phytophthora infestans* on cut surfaces of potato tubers. Hyphae from zoospores penetrated resistant cell walls within a few hours from inoculation (Tomiyama, 1967) and hyphal growth was noted to stop inside resistant cells several hours later. Another series of observations concerned the stage of the infection process when four different genes for resistance in wheat to *Erysiphe graminis* f. sp. *tritici* expressed themselves. Working with nearly synchronously developing inocula under controlled environmental conditions, Slesinski & Ellingboe (1969) found that none of the genes had any effect in the first 12 hours when appressoria formed, nor during the next few hours when primary haustoria formed inside epidermal cells. Hyphal growth stopped at later stages, as described in Chapter 3, again indicating that critical events occurred after penetration of cell walls.

CONCLUSION

Failure of parasitism may occur in some host–parasite interactions during attempted penetration of cuticle and epidermal walls as with *Colletotrichum gloeosporioides* in the poplar. These penetrations are very difficult to observe and to measure quantitatively. However, in most other studies of early stages of infection, failure of parasitism occurs at a later stage, at least after entry into the first cell penetrated. The best-known defence mechanisms of plants operate, therefore, after penetration into the cells, but it must be repeated that little is known of the fate of many potential parasites in nature. Clearly, many interactions are possible during the earliest stages of attempted parasitism as indicated in this chapter. Further research might reveal these interactions to be more important in the general defence of plants than present evidence indicates.

CHAPTER 3

Cytological Changes in Host and Parasite after Infection

It must be clear from the previous chapter that the fate of most potential parasites is decided after they have entered their host plants. It is important, therefore, to know the location of parasites within and between cells, the stages during attempted infection when parasitism succeeds or fails, and the associated responses of host cells as revealed by light and electron microscopy. Based on this knowledge, the significance of the numerous physiological and biochemical changes in infected plants to the processes of resistance and susceptibility can be assessed.

THE CELLULAR LOCATION OF PARASITES IN HOST TISSUES

There have been a limited number of quantitative studies on related cytological changes in host and parasite during the critical stages when resistance is expressed or parasitism is established. Before considering these studies which are confined to a few much analysed host–parasite relationships, it is valuable to review the different ecological niches presented by plants and the ways in which they are exploited by specialized microorganisms. In doing so, it is well to emphasize the types of physiological interaction which might be anticipated between the cells of parasite and host, and to appreciate the rarity of direct contact of the protoplasts of the two organisms. The parasitic protoplast is usually separated by its cell wall from the host, and in many cases a host cell wall also separates the two protoplasts.

Perhaps one of the most unusual niches is that between the cuticle and the outer epidermal wall of leaves, which is exploited by the mycelium of *Venturia inaequalis* and *Diplocarpon rosae* in the apple *Malus sylvestris* and *Rosa* spp., respectively. Successful growth of mycelium in this site is followed some time later by the deterioration and discoloration of the underlying epidermal cells, which are not directly contacted by the fungus (Nusbaum & Keitt, 1938; Preece, 1963). Envisaged physiological interactions with the host deciding the fate of parasitism might be with components of cuticle and epidermal wall or, more remotely, with the epidermal cells beneath.

The outer surface of many plants provides a site for the growth of powdery mildew fungi in the genus *Erysiphe*. Contact with epidermal cells is by means of haustoria, which are put through the cuticle and epidermal wall. The protoplasmic membranes of these cells are invaginated by the haustoria which become large and develop finger-like lobes. The host protoplasts are always separated from the fungal protoplast, not only by a fungal wall but also by a sheath which may be a product of the fungus and host or of the host alone (Bracker & Littlefield, 1973). Compatibility of haustoria and host protoplasts must be a prerequisite for the development of powdery mildews, which are biotrophic parasites. Not only can many possible interchanges be envisaged between the haustorium and the host cell, but also there is evidence of a metabolic interaction between host and parasite in the first hours of mildew development, before the host is penetrated. Germinating conidia of *Erysiphe*, but not of other organisms, affect the physiology of underlying wheat leaves and cause stomatal closure (Martin, Stuckey, Safir & Ellingboe, 1975). Some volatile or diffusible fungal metabolite must pass into host cells at the very earliest stages of attempted parasitism.

Many parasitic organisms exploit the intercellular spaces between mesophyll cells of leaves as a habitat for growth. Thus, most rust fungi grow from substomatal cavities by means of intercellular hyphae which push haustoria into contacted mesophyll cells. Suggestions that the young haustorium lacked a fungal wall (Ehrlich & Ehrlich, 1963) and that direct contact between rust and host protoplast occurred have been discounted by recent examinations of the fine structure of the parasite during cell penetration (Heath & Heath, 1971; Bracker & Littlefield, 1973). Compatibility between the walled haustorium, the surrounding sheath and the host plasma membrane again seems essential for successful parasitism. Interactions are known to occur between intercellular hyphae and host cells before wall penetration and are quite common between young haustoria and protoplasts.

Intercellular spaces in leaf mesophyll are the major sites for the multiplication of many parasitic bacteria in the genera *Pseudomonas* and *Xanthomonas*. Substantial population growth occurs within hours of inoculation (Ercolani & Crosse, 1966) when major differences between the multiplication rates of avirulent and virulent organisms become apparent. Host cells are not entered during this important period, and exchanges with the bacterial cells must occur by diffusion of substances through walls. Even when very low concentrations of avirulent bacteria are introduced into intercellular spaces, host cells die within a few hours (Turner & Novacky, 1974), whereas virulent bacteria maintain a compatible relationship with host cells for much·longer periods.

Some parasitic fungi undergo most of their development inside living host cells. Thus, *Colletotrichum lindemuthianum* penetrates through the cuticle and produces large hyphae inside successive epidermal and cortical cells of susceptible bean stems, petioles and veins (Skipp & Deverall, 1972). Only the hyphal wall separates the fungal protoplast from the host plasma membrane, which is pushed aside as the hyphae grow through the cells (Mercer, Wood & Greenwood, 1974). Harmony between host and parasite must depend on the compatibility of the host protoplast with the fungal wall and with substances secreted through it.

Direct contact of host and parasite protoplasts occurs in the relationships between plants in the Cruciferae and the plasmodial parasites *Olpidium brassicae* (Lesemann & Fuchs, 1970) and *Plasmodiophora brassicae* (Williams & McNabola, 1970). Very soon after penetration from encysted zoospores of *P. brassicae*, the parasitic amoebal cell in compatible host cells becomes bounded by a multi-layer of membranes. Williams, Aist & Bhattacharya (1973) have speculated that the fate of parasitism might be decided within moments of introduction of the amoeba into the cell.

Successful development of vascular parasites such as *Fusarium oxysporum* and *Verticillium* spp. depends upon their ability to penetrate and grow through root tissues until they reach the lumen of xylem vessels. They colonize these non-living vessels by mycelial growth and liberation of conidia which are carried in the transpiration stream through perforation plates into adjacent vessels. Interactions with plant tissues affecting the success of parasitism could occur at any stage before entry into xylem vessels and, thereafter, with components of vessel walls and possibly with the live parenchyma cells adjacent to vessels (Beckman, 1971; Talboys, 1972).

Those organisms, which are necrotrophic for most phases of their parasitic development, probably kill host cells by toxic secretions and thus rarely contact living cells. Fungal and bacterial parasites which cause soft-rot diseases in parenchyma of tubers and fruit are wound entrants and then producers of enzymes which separate and kill host cells (Tribe, 1955; Garibaldi & Bateman, 1970; Stephens & Wood, 1975). In return, metabolites may diffuse from responding live host cells through the dead tissue to affect the parasites. Similar secretions might also emanate from host cells ahead of the necrotrophic leaf-spot fungi which destroy cell walls and protoplasts in their immediate vicinity. Thus, much less immediate interactions between the necrotrophic parasites and their hosts might be anticipated, in comparison with the many parasites which are biotrophic for much of their lives.

CESSATION OF GROWTH OF PARASITES IN RESISTANT PLANTS

Answers to the questions of where and when parasites stop growing in resistant plants have been provided in varying detail for different types of parasite. Greatest precision has been achieved by Ellingboe and his colleagues, in their analysis of the fate of the powdery mildew fungus *Erysiphe graminis* f. sp. *tritici* in wheat cultivars bearing different single genes for resistance. As discussed by Ellingboe (1968, 1972) this work was organized so that the effects of genes conferring resistance or susceptibility in the plant and avirulence or virulence in the fungi could be discerned under controlled environmental conditions. These conditions were not only controlled for the growth and infection of the wheat but also for the production of the conidia used as inocula. Thus, successive experiments could be conducted to relate physiological changes to observed morphological events. Furthermore, within any experiment, the development of all conidia was almost synchronous. The results showed that the different genes controlling the host–fungus interaction had no effect in the first 12 hours, when conidia germinated and produced appressoria (Slesinski & Ellingboe, 1969; Stuckey &

21

Ellingboe, 1974). Primary haustoria were produced similarly in the different interactions. Some of the genes for resistance caused the parasite to stop growing within the next 18 hours, but other genes acted less effectively or at much later stages. Cessation of hyphal growth required the presence of complementary genes for avirulence in the fungus and resistance in the host, as predicted from the gene-for-gene hypothesis (Slesinski & Ellingboe, 1970). Thus the parasite–host genotype $P4/Pm4$ proved to be the most effective, causing the collapse of about 95% of the secondary hyphae elongating from appressoria 22 hours from inoculation, when host cytoplasm discoloured around the haustoria. Next in effectiveness was the $P1/Pm1$ genotype, which caused about 80% of secondary hyphae to cease growth 26 hours from inoculation, but without associated discoloration in penetrated host cells. Less effective was the $P3a/Pm3a$ genotype which allowed at least 30% of secondary hyphae to continue to elongate after 26 hours. The $P2/Pm2$ genotype was effective only after several days had passed, when host cells died and only a limited crop of conidiophores was produced. These elegant studies demonstrate that different genotypes for low infection type act in different ways and at different stages of the host–parasite interaction. Molecular explanations of the different phenomena are not yet available, but the cytological studies have been followed by analyses of the ways in which different parasite and host genes affect the transfer of ^{35}S from the plant to the mildew mycelium (Slesinski & Ellingboe, 1971; Hsu & Ellingboe, 1972).

It is more difficult to follow the progress of parasites which penetrate deeply in plant tissues, unlike the members of the genus *Erysiphe*. Particularly difficult problems are presented by the rust fungi because it is necessary not only to record the growth of intercellular hyphae but also to note the stages of hyphal growth when contact is made, via haustoria, with cells within leaves and stems. The possibility exists that haustorial penetration might occur more frequently in some host–parasite interactions than in others, and that it might be unrelated to the spread of intercellular hyphae. The earliest investigators (Ward, 1905; Stakman, 1915) observed that a very common result of the contact of rust hyphae with cells of resistant hosts was the rapid death of the cells and cessation of further growth of the hyphae. The rapid death of resistant cells was termed *hypersensitivity* by

Stakman. Although Stakman made no implications concerning the sequence and nature of physiological factors which resulted in the failure of further hyphal growth, it is clear that any analysis of the time course of events in resistant plants must record the hypersensitivity of host cells, in addition to the penetration of these cells from intercellular hyphae. Hypersensitivity of plant cells has been recognized to be commonly associated with resistance of plants to many fungi (Müller, 1959) and bacteria (Klement & Goodman, 1967). The processes involved in the initiation of hypersensitivity and in the limitation of potential parasites are matters of lively debate and investigation, as discussed in later parts of this book.

The time in the infection process when the development of rust hyphae stops in resistant plants has been estimated in several ways. Ogle & Brown (1971) cleared and stained whole leaves of cereals and measured areas of mycelial colonies of *Puccinia graminis* f. sp. *tritici* by microscopic examination through surfaces. They concluded that rust growth stopped within 48 hours of inoculation in highly resistant wheat cultivars and in other cereals. They also observed that the area of these limited colonies was exceeded by an area of necrotic host cells. Skipp & Samborski (1974) followed the progress of one race of this rust in isogenic lines of wheat, differing only in the presence or absence of the Sr_6 gene governing resistance to this race at 20 °C. In general, they confirmed that most rust hyphae grew little more after the second day in the presence of the gene for resistance. However, they observed that the first haustoria were produced in a similar way in resistant and susceptible lines 16 to 20 hours after inoculation. Differences between the lines became apparent several hours later, when some hypersensitivity occurred in resistant mesophyll cells. Differences in hyphal growth became noticeable 36 hours after inoculation and substantial 24 hours later. Although growth continued freely in the susceptible line, growth was restricted in resistant tissue where necrosis was visible. The problems presented for an analysis of the sequences of the physiological processes that control these events are emphasized by their observations that infection sites behaved differently, even in a resistant leaf. For example, epidermal cells never underwent hypersensitivity although penetrated by haustoria; some mesophyll cells became necrotic rapidly, possibly without haustorial penetration; and other mesophyll cells did not become necrotic until several neighbouring cells contained haustoria.

Another way of attempting to decide when the process of resistance is complete, following rust infections, is through manipulation of the sensitivity of expression of the Sr_6 gene to temperature. This gene confers low infection type when plants are grown and incubated at 20 °C but high infection types at 26 °C. Antonelli & Daly (1966) found that transfers of plants from one of these temperatures to the other, at any stage within three days from inoculation, caused the development of the infection type normally characteristic of the final temperature. This suggested that metabolic processes regulating the infection type are not effective until three days after inoculation. Transfers at stages after this period caused the development of intermediate infection types. When Skipp & Samborski (1974) did similar experiments and observed histological changes during the three-day period, they noted that host cells responded within 10 to 20 hours to a change in temperature. Thus, cellular necrosis and accompanying metabolic changes can occur at very early stages in the infection of resistant leaves. The results of these experiments and observations show that rust hyphae are not killed in necrotic tissue. Although haustoria and subtending hyphae die, growth can resume from a more remote mycelium when appropriate changes in temperature are made. Skipp & Samborski (1974) also deduced that the temperature dependence of the expression of the Sr_6 gene probably resided in the wheat plant, because they found that the temperature at which plants were grown before inoculation influenced the sensitivity of cells to the rust. Some cells in plants grown at 20 °C became necrotic when inoculated and incubated at 25 °C, unlike those in plants kept at 25 °C throughout. Knowledge of these visible cellular reactions should aid interpretation of experiments on metabolic changes, as resistance conferred by the Sr_6 gene is expressed. The relationship between hypersensitivity and rust development will be considered again later.

Although not the natural sites for infection by fungi, the interiors of legume pods and the fruit of the pepper *Capsicum frutescens* have often been used to study physiological interactions with different fungi, for the reasons explained in Chapter 5. It is therefore useful to note here the types of cytological change which take place in pepper fruit cavities following infection by some virulent and avirulent parasites (Jones, Graham & Ward, 1974, 1975a, b). Introduction of zoospores of the virulent fungus *Phytophthora capsici* caused rapid responses in the underlying cells which produced many ribosomes and showed nuclear changes within four hours. Chloroplasts were beginning to degenerate at this time, and two hours later the cytoplasms of these cells were disorganized as the hyphae penetrated deeper into the tissue. Newly approached cells also became ribosome-dense before degenerating after penetration. Thus, this virulent parasite seemed to reactivate cells before killing them and growing on apparently unchecked. Greater incompatibility with pepper cells was shown by *P. infestans* which killed the first cells penetrated within four hours. The fungus spread into two or three layers of cells which died after penetration. Adjacent uninvaded cells became ribosome-dense, and the fungus was limited within 36 hours. Hyphal death and disintegration were observed to follow host cell death. The soft-fruit parasite *Monilinia fructicola* was highly incompatible with pepper because it only succeeded in penetrating a few cells, yet caused widespread host necrosis ahead of the hyphae. This study shows that two types of cellular response could be distinguished, namely rapid necrosis and slower death often preceded by an apparent reactivation of the ribosomal apparatus for synthesis. *Phytophthora capsici*, unlike the other fungi studied, was able to colonize pepper fruit despite its incompatible relationship with penetrated cells.

THE ASSOCIATION OF HYPERSENSITIVITY
WITH RESISTANCE

A recurrent theme in many investigations on resistance of plants to rust parasites and many other parasites is the close association between hypersensitivity and resistance. Thus it is quite likely that resistance results from the hypersensitive death of the host cell, perhaps because the haustorium within the host cell is starved when the protoplast dies, or even killed by toxic substances in the dead cell. However, the idea that hypersensitivity is a cause of resistance has been challenged in several ways in recent years.

One way results from extensive analyses of the growth of black stem rust *Puccinia graminis* f. sp. *tritici* in a range of wheat cultivars by Brown, Shipton & White (1966) and by Ogle & Brown (1971). The essence of the argument was that three categories of host reaction could be recognized on the basis of rust growth and host necrosis, namely resistant, susceptible and intermediate. In the resistant reaction, mentioned in the previous section, the small rust colonies were exceeded by areas of necrotic tissue. In the susceptible reaction, rust colonies became very large and host necrosis was negligible. These two observations were consistent with the idea that necrosis caused limitation of rust growth. However, in the intermediate reactions, moderate colony areas exceeded very variable areas of host necrosis; this implied that the necrosis had not prevented rust growth and was the basis for the suggestion that resistance was not caused by death of the host cells.

It would be most interesting if some of these intermediate reactions could be examined for the sequence of interactions between hyphae, haustoria and host cells in the first stages of infection, as done by Skipp & Samborksi (1974) for the resistant reaction mediated by the Sr_6 gene. A low sensitivity of cells to haustorial penetration in some intermediate reactions might have resulted in late necrosis and permitted considerable hyphal growth compared with the resistant types. One of the intermediate reactions was in the cultivar Einkhorn, which had also been regarded as distinctive by Stakman (1915) because rust growth was limited despite the absence of any obvious cellular reaction. Clearly, in this cultivar, a process quite unrelated to cellular necrosis limits rust development. The observations on the remaining intermediate reactions indicate nicely the need for further histological investigations. They also draw attention to the deficiencies in knowledge of fungal growth and host response in plant cultivars intermediate in reaction to other parasites, particularly in cultivars with so-called 'field' resistance.

The role of hypersensitivity in resistance of wheat to *Puccinia graminis* f. sp. *tritici* has also been brought into question by observations on host cell death, as measured by fluorescence microscopy, in plants possessing the Sr_6 gene (Mayama, Daly, Rehfeld & Daly, 1975). The number of fluorescent sites per unit area of leaf increased in the same way irrespective of whether a low infection type was being produced at 20 °C, or a high infection type at 26 °C. Thus, substantial uredial formation occurred at 26 °C despite the occurrence of the same pattern of cell death in response to haustorial penetration as at 20 °C. Assuming that the technique of detecting fluorescent cells records cells that died in response to infection, this study implies either that cell death *per se* has no effect on rust development or that some rust-limiting process linked to cell death does not function at 26 °C.

Another way in which hypersensitivity as a cause of resistance is questioned, arises from experiments performed by Kiraly, Barna & Ersek (1972). They used chemical agents to stop fungal growth in susceptible plants, and then observed responses in host cells around the infection hyphae. For example, chloramphenicol prevented the growth of a compatible race of *Phytophthora infestans* in potato tubers which, in turn, caused hypersensitivity of the tuber cells. Similarly, necrosis was induced in wheat leaves inoculated with a compatible race of *Puccinia graminis*, and in bean leaves inoculated with *Uromyces phaseoli* after treatment with fungal inhibitors. The implications of this work are that fungal growth in resistant plants is stopped by an unknown factor, and that the inhibited fungus then releases a toxic substance which kills the host cells. The latter possibility was supported by evidence that a killed mycelium of *Phytophthora infestans* released a fluid *in vitro* which induced hypersensitivity when introduced into potato tissue. These experiments lead to the conclusion that hypersensitivity can be a consequence and not a cause of resistance.

The apparent force of these arguments is much decreased by observations that, in a number of host–parasite interactions, fungal hyphae continue to grow, albeit increasingly more slowly, in plant cells after these cells have undergone hypersensitivity. For example, Tomiyama (1955) observed the growth of incompatible hyphae of *Phytophthora infestans* inside potato cells for several hours after the cells had become necrotic. Growth eventually stopped before the hyphae passed into a second cell. These observations are supported by a study of the ultrastructure of the interaction between this fungus and potato leaves (Shimony & Friend, 1975). Penetrated and adjacent epidermal and mesophyll cells of resistant leaves died within 9 to 12 hours of inoculation, but the hyphae were not killed until several hours later. Skipp & Deverall (1972) measured hyphal lengths in 'marked' cells in excised bean tissue mounted on microscope slides through an 18-hour period, when the cells were beginning to undergo hypersensitivity to an incompatible race of *Colletotrichum lindemuthianum*. Growth rates of hyphae inside resistant cells before symptoms were detected ranged up to 9 μm/hour, but were lower in cells where cytoplasm was becoming granular as part of the necrotic reaction. However, slow hyphal growth was detected in cells which had become pale brown in colour. This was supported by other observations using whole hypocotyls which indicated that some hyphae remained alive and grew in brown, apparently dead, cells, although they did not grow out of these cells. This suggests that a progressive inhibition of fungal growth follows hypersensitivity of the host cell. This traditional view of the sequence of events in hypersensitive cells is supported by the report by Maclean, Sargent, Tommerup & Ingram (1974) that the membranes and cytoplasm of resistant cells of the lettuce *Lactuca sativa*, viewed beneath the electron microscope, became severely disrupted four hours after penetration by the downy mildew *Bremia lactucae*. At this time, fungal cytoplasm appeared normal. The fungus continued to grow in the cell for another 12 hours before dying. The details of observed changes in hypersensitive cells do not, therefore, support the conclusions, based on the experiments performed by Kiraly *et al.* (1972), that hypersensitivity is a consequence of resistance having been expressed earlier by unknown factors.

OTHER CELLULAR REACTIONS OBSERVED TO BE ASSOCIATED WITH RESISTANCE

Hypersensitivity is not the only visible cellular process associated with resistance. Reports have shown that other reactions can occur, and a good example is that of the enclosure of rust haustoria by a particularly dense sheath. Electron microscopy on sections of cells of the cowpea *Vigna sinensis* resistant to the cowpea rust *Uromyces phaseoli* f. sp. *vignae* reveal at least two fates of incompatible haustoria, namely collapse in hypersensitive cells and encapsulation by very thick sheaths in otherwise unaffected cells (Heath & Heath, 1971). Similar studies by Heath (1974) show that a diversity of cellular responses to cowpea rust can occur in cowpea cultivars and other plant species, some apparently deterring haustorial formation and others affecting the haustorium in the penetrated cell. In addition to hypersensitivity and encapsulation, two other phenomena visible by electron microscopy were deposition of osmiophilic material on the inside of a cell wall opposite to an intercellular hypha, and poor adhesion of haustorial mother-cells and plant cell walls. As a result of this survey, Heath (1974) envisaged that parasitism by a rust can be regulated by a series of 'switching-points', and that the result at each determines whether the fungus can progress to the next. The greater the compatibility between parasite and host, the further the rust progressed so that in a susceptible plant, no signs of deleterious reactions to the rust could be discerned.

CONCLUSION

Recent research reveals that growth of parasites can stop at a number of stages inside the tissues of resistant plants, and different types of processes are likely to be involved at these different stages. Most studies have concerned the more complete forms of resistance where infection stops within one or two days before the parasite has progressed far. Although there are some contrary indications, processes accompanying hypersensitive host cell death often seem to be related to the expression of rapid resistance to incompatible races, but other imperfections in the host–parasite interface can occur. Relatively little information is available on the progress of parasites inside plants which are intermediate in reaction type or which possess so-called 'field' resistance. However, one excellent example was provided of the delayed expression of a gene-for-gene interaction causing incompatibility and low infection type after a period of compatible growth of powdery mildew in wheat. This suggests that genetically determined incompatibility can occur at any stage in the ontogeny of the interaction between host and parasite.

CHAPTER 4

Cross-Protection and Induced Resistance

Cross-protection is the phenomenon whereby a plant is protected from infection by earlier or simultaneous exposure to
another organism. This chapter will be concerned with protection against fungal and bacterial parasites as a result of infection
by related and unrelated fungi and bacteria and of local lesion
formation by viruses. It will not be concerned with the interference of the development of viruses by related or unrelated
viruses (Matthews, 1970).

For a long time people have been interested in the idea that
plants, like animals, can acquire physiological immunity to
pathogens. Research, mainly in Europe.in the late part of the
nineteenth century and in the early decades of this century, was
inspired by observations which suggested that some perennial
plants were less severely affected by a disease following earlier
infection. For example, the varied development of powdery
mildew infections on the foliage of oak trees *Quercus* spp. in
successive seasons had intrigued early investigators. Experiments had also been performed which seemed to show that
herbaceous plants became more resistant to infection after the
plants or the soil bearing them were sprayed with the extracts of
fungal cultures. Numerous observations and claims of this type
were reviewed by Chester (1933) who criticized the lack of
sufficient replication for proper appraisal of the experiments
and the frequent inadequacy of controls. Sources of error which
had rarely been eliminated before drawing conclusions were
losses of virulence in the parasites during experiments, differences in environmental conditions on successive occasions for
assessing resistance, and natural increases in resistance as plants
aged during the periods of supposed acquisition of resistance.
From all of this work, there remain possibilities, requiring
substantiation, that trees or new branches are less susceptible
after infection of earlier growth and that antibiotics released by

protecting organisms might be responsible for some of the effects.

The early work with the most convincing outcome followed that of Bernard on the interactions between germinating orchid seed and fungi, leading to the development of functional seedlings with mycorrhizal roots. Some fungi destroyed the seed, some did not survive attempted infection of the seed and others entered into association with the roots of the seedling to form mycorrhiza. Using excised orchid embryos, Bernard (1909) found that one of these types of fungi penetrated several layers of cells before it stopped growing and disintegrated. Attempts to re-infect the embryo with a normally destructive fungus were then unsuccessful, suggesting acquisition of resistance in the embryo. Further experiments gave an indication of the involvement of antifungal compounds arising from the orchid tissue. Bernard (1911) plated a piece of surface sterilized tuber of the orchid *Loroglossum hircinum* a short distance from the pathogen *Rhizoctonia repens* on solid culture medium. After beginning to grow in all directions, the fungus became inhibited as it approached the orchid tissue but before it reached the tuber. Although these tuber fragments were fungistatic, it was not possible to detect any activity in extracts of crushed fresh tubers. Fragments heated at 55 °C for 35 minutes were also inactive. Nobécourt (1923) confirmed these results and found that tuber fragments which had been frozen and thawed or exposed to chloroform vapour also did not inhibit the growth of the fungus. He thought it unlikely that such diverse treatments as exposure to heat, cold and chloroform would destroy a pre-existing fungistat in the tissue. The fact that all three treatments killed the tissue led Nobécourt to suggest that they prevented the tissues from synthesizing the fungistat in response to substances diffusing from the fungus. Magrou (1924) disputed this interpretation, and showed that sterilized pieces of tuber incubated alone on medium for two weeks exuded a fungistat into the medium. Although it was clear therefore that orchid tissue liberated an antifungal material under some circumstances, it was uncertain whether Magrou had detected the same material as Nobécourt and Bernard. If it was the same material, there was no indication of its quantity in the two types of experiment. Furthermore, considerable doubt existed about the nature of the stimulus required for its formation and/or liberation; cutting the orchid tissue might have been the essential stimulus. These important questions were unresolved in 1933 and remained so until the work of Gaümann and his associates, which led to the discovery of a number of antifungal compounds produced by different orchid tubers in response to infection, as discussed in Chapter 5.

The next important and extensive investigation on acquired resistance was that of Müller & Börger (1941) on the interactions of potato varieties with different fungi, particularly with races of *Phytophthora infestans*. An essential basis for these experiments was Müller's earlier work on the selection of potato varieties resistant to *P. infestans*, which also led to the detection and collection of virulent and avirulent races of this fungus.

Müller & Börger showed that treatment of cut surfaces of potato tubers with an avirulent race prevented the development of a virulent race inoculated a day later. Cut tubers, aged for several days in humid chambers in the absence of the avirulent race, retained their susceptibility to the virulent race. The protection was total when one or two days elapsed between successive inoculations but was recognizable only by weaker mycelial growth when the intervals between inoculations were as short as one hour. The protection was also localized, being confined to the area of the tuber surface treated with the avirulent race, and was apparently non-specific, being effective against an isolate of *Fusarium* normally pathogenic towards potato tubers. Thus, by means of these experiments, localized non-specific protection was demonstrated in potato tubers.

There are now several recent reports of systemic cross-protection (e.g. Kuć, Shockley & Kearney, 1975) to set beside the many reports of localized and non-specific protection against both fungal and bacterial pathogens. Thus, for example, avirulent rust fungi will protect against virulent rusts on leaves and stems of several plants (Littlefield, 1969; Kochman & Brown, 1975). Avirulent races of *Pseudomonas* spp. will protect against virulent races in intercellular spaces of plants (Averre & Kelman, 1964). Avirulent isolates of *Fusarium oxysporum* protected against virulent isolates in experiments with seedlings in test tubes, under conditions which would seem greatly to have favoured the development of a general parenchymatous infection by the pathogen, even if not a characteristic vascular infection (Davis, 1967). Selected reports will be referred to in the remaining parts of this chapter where they make a contribution to understanding the different possible modes of action of protection.

MODES OF ACTION OF CROSS-PROTECTION

There are two distinct ways in which protection may be brought about. Firstly, the protectant organism could act directly on the normally pathogenic organism by physical impedance or chemical antagonism. Secondly, the protectant could activate a physiological change in the host plant so that it becomes resistant. Such induced resistance would be of greatest consequence to the main theme of this book, but it is first desirable to consider the evidence that induced resistance can be distinguished from direct interference.

DIRECT ACTION OF THE PROTECTANT
AGAINST THE PATHOGEN

One of the most likely ways in which a protectant organism can be effective is by blocking the stomatal sites of entry for the normal pathogen. The most important stomatal entrants for which cross-protection has been shown are the rust fungi. After demonstrating localized protection of wheat leaves against *Puccinia recondita* by treatment with the oat rust *P. coronata* f. sp. *avenae*, Johnston & Huffman (1958) suggested that the stomata may have been blocked by the protectant. However, in the course of experiments in which he achieved marked localized and non-specific protection against the flax rust *Melampsora lini*, Littlefield (1969) calculated that insufficient stomatal apertures would have been blocked by the low concentrations of uredospores which were effective in causing protection. Evidence that cross-protection between rusts is sometimes caused by blocking sites of entry was provided by Kochman & Brown (1975) who observed some occlusion of stomata on oat leaves by the appressoria of wheat rusts used as protectant in their experiments.

Johnston & Huffman (1958) also suggested that self-inhibitors released by the uredospores of oat rust may have been effective in preventing infection by wheat rust. It is well known that many uredospores release inhibitors of germination into surrounding fluids (Allen, 1955). Yarwood (1954) also showed that volatile emanations from rusted bean and *Antirrhinum* leaves inhibited germination of uredospores and impaired development of other rust infections. Yarwood (1956) compared the efficacy of uredospores with fungicides as protectants against rusts. Thus he showed that 1–3 mg dry weight of uredospores of bean rust per dm^2 of the leaf of the sunflower *Helianthus annuus* gave 50% control of uredial production by the sunflower rust *Puccinia helianthi*. Conversely, 4 mg of uredospores of sunflower rust gave similar control of bean rust. The effect was considered to be caused by self-inhibitors. However, unless self-inhibition is much stronger between uredospores of different species than between those of the same species, it is surprising that there was no self-limiting effect of increasing the dosage of a pathogenic rust alone to 10 mg/dm^2. This inoculum gave rise to so many uredia that it was impossible to count them. Thus there is reason to doubt that inhibition of spore germination by spore products was the cause of protection in the earlier experiments. Several self-inhibitors have now been characterized (Macko, Staples, Allen & Renwick, 1971; Macko, Staples, Renwick & Pirone, 1972); bean rust, sunflower rust and some strains of wheat rust uredospores have been shown to be sensitive, and flax rust uredospores insensitive, to methyl-3,4-dimethoxycinnamate released from uredospores of the sunflower and *Antirrhinum* rusts. It should be possible now to assess the extent to which self-inhibitors could cause cross-protection between rusts. Littlefield (1969) observed that germination of flax rust was unimpaired by the protecting rust on flax leaves, thus showing clearly that self-inhibitors were not involved in his experiments.

Direct chemical antagonism by a protective organism is likely to be involved in the prevention of crown-gall development *Agrobacterium tumefaciens* by treatment of susceptible plants with the non-pathogenic *A. radiobacter*. This treatment is the basis of a practical control of crown-gall disease in South Australia (Kerr, 1972; New & Kerr, 1972). The mode of action is thought to be through the secretion of a specific antibiotic, a bacteriocin, by the protectant bacterium. Kerr & Htay (1974) showed that many strains of *A. tumefaciens* were prevented from causing crown-gall when they were inoculated into wounds in tomato stems in 1:1 mixtures with strain 84 of *A. radiobacter*. Growth of all these strains of *A. tumefaciens* was inhibited on nutrient agar plates around spots where strain 84 had previously been allowed to grow for two days before being killed with chloroform. Thus, it seems likely that protection in tomato is caused by a similar inhibition of growth of *A. tumefaciens*. Diminished growth of one pathogenic strain in tomato was demonstrated, but it is desirable to show that the bacteriocin is produced *in vivo* by the protectant strain. An alternative hypothesis to explain protection has been suggested, namely competition for sites for bacterial attachment or action within host tissues (Lippincott & Lippincott, 1969).

Both direct chemical antagonism by bacteriocins, and exclusion from infection sites, were considered unlikely explanations by Garrett & Crosse (1975) for the similar suppressive effect of plum strains of *Pseudomonas morsprunorum* and some other *Pseudomonas* spp. on the development of leaf-scar infections and cankers in the cherry *Prunus avium* caused by the cherry strains of this organism (Crosse & Garrett, 1970). There was no relationship between the suppressive action of *Pseudomonas* spp. in cherry and their antagonistic effects on the cherry strain by lysogeny or bacteriocin production *in vitro*. Furthermore, related *Pseudomonas* spp. which might be expected to compete for infection sites in cherry had no effect on canker development by the cherry strain. The common feature of the strains and species of *Pseudomonas* which suppressed canker development was their ability to induce hypersensitivity in non-host plants. Thus, Garrett & Crosse (1975) suggested that avirulent strains inhibited virulent strains as a result of induction of hypersensitivity in the host, in agreement with conclusions by Averre & Kelman (1964) about a similar phenomenon revealed in studies with *P. solanacearum*.

INDUCED CHANGES IN THE HOST PLANT

There are several types of evidence that cross-protection can result from a change in the physiology of the host plant. The first arises from the studies of Müller & Börger (1941) on cross-protection in tubers against potato blight. When a thin layer of tuber tissue was excised beneath a surface treated a day earlier with an avirulent race of *Phytophthora infestans*, the underlying tissue retained its resistance to a virulent race. This eliminated physical hindrance as a cause of the protection but left the possibility that an antibiotic had diffused from the avirulent race. However, the occurrence of direct chemical antagonism between the avirulent race and the virulent race was eliminated by the results of an experiment in which a mixture of the races was inoculated into a potato cultivar susceptible to both races. A large crop of sporangia was produced, and a dilution series of these sporangia was inoculated onto differential cultivars of potato. By this means, both races of *P. infestans* were recognized in equal amounts among the sporangia. The avirulent race therefore exerted its action on the virulent race only in resistant host tissue, implying that it activated a host process to become effective against the normally virulent race.

A second type of evidence comes from direct observations of the behaviour of protectant and pathogenic spores and germ-tubes during the infection process. Littlefield (1969) working with flax rust and Skipp & Deverall (1973) with bean anthracnose saw no effect of avirulent spores on normal germination and penetration by virulent spores. Their conclusion was that protection was achieved after entry of both types of fungus into the host plant, and that there was no direct interference between the fungi.

Striking evidence of a change in host physiology leading to resistance to a fungal pathogen is provided by work with tobacco and a strain of tobacco mosaic virus (TMV) which caused local lesion formation. Local lesion formation on one side of a tobacco leaf made the opposite side of the leaf much more resistant to the fungal pathogen *Thielaviopsis basicola* (Hecht & Bateman, 1964). Strains of virus which did not cause local lesions had no effect on fungal development. This is not the only report of remote cross-protection in tobacco, because field observations by Pont (1959) prompted Cruickshank & Mandryk (1960) to investigate the decrease of foliar infection by *Peronospora tabacina* as a result of earlier stem infection by the same fungus. When conidia were injected into the stub of a cut petiole near the base of tobacco stems, the upper foliage became completely resistant to sporulation of the downy mildew sprayed onto the leaves under optimal test conditions four weeks later. This remarkable result is put into better perspective when it is realized that the stem-infected plants became prematurely senescent compared with the water-injected controls. The leaves on the protected plants showed signs of senescence, and the plants matured and flowered two weeks before the controls. Thus it is not clear whether the resistance of the leaves was caused indirectly by their premature senescence. These varied phenomena in tobacco are open to several other interpretations such as competition between infection sites and the formation of systemically mobile antifungal substances. The value of tobacco as a worthwhile subject for deeper investigation of induced resistance is increased still further by the discoveries of Lozano & Sequeira (1970) of its response to heat-killed bacteria to be discussed later.

Remote cross-protection, as distinct from the often demonstrated localized form, has also been claimed in recent research with anthracnose diseases of bean and cucurbits and with bacterial diseases of pome fruit. Avirulent races of *Colletotrichum lindemuthianum* caused sites 5 mm away on etiolated bean hypocotyls to become resistant to virulent races (Elliston, Kuć & Williams, 1971). This implies that some factor is translocated in etiolated tissue to induce a change at a distance of numerous cell-widths, although it must be noted that Skipp & Deverall (1973) were unable to achieve similar results on green hypocotyls. More remarkable, is the demonstration that infection of the first leaf of the cucumber *Cucumis sativus* with *Colletotrichum lagenarium* renders the younger foliage resistant to infection by the same fungus when this is applied again one or more weeks later (Kuć, Shockley & Kearney, 1975). Direct interference of one inoculum with the next is clearly impossible and induction of premature senescence of the plant seems unlikely in the time involved and from the appearance of the plants. Thus, the activation of an uncharacterized process of resistance throughout the plant is implied. Also largely unexplained is the means by which application of preparations containing DNA from the bacterium *Erwinia amylovora*, to roots or cut shoots of seedlings of the pear *Pyrus communis*, protects these seedlings from subsequent infection by the bacterium (McIntyre, Kuć & Williams, 1975).

THE NATURE OF THE INDUCED CHANGE IN THE HOST PLANT

A simple explanation of the types of change in host plants which increase their resistance would be the diffusion of anti-microbial compounds from the host tissues at the sites of inoculation with the protectant organism. In fact, the original concept of phytoalexins was that they were produced when cells became

37

necrotic in hypersensitive responses to avirulent fungi (Müller, 1958). This idea had arisen during earlier work on cross-protection in potato (Müller & Börger, 1941); a hypothetical antifungal principle was conceived to diffuse out from hypersensitive cells to prevent development of virulent races of the potato blight fungus. The formation of phytoalexins in and around necrotic cells in a number of plant families has now been demonstrated clearly, as discussed in Chapter 5. However, there is no evidence that any of the known phytoalexins diffuse from sites of formation, although there is a demonstration of phytoalexin formation in live cells next to dead cells, as discussed in Chapter 5. There is no reason as yet, therefore, to assume that cross-protection is caused by diffusion of phytoalexins over the distance of even several cell widths.

As a matter of speculation, it is quite possible that phytoalexins with appropriate solubility might be able to diffuse through necrotic cells, where barriers to permeability and movement of molecules have been destroyed. Furthermore, systemic fungicides must be able to move for considerable distances in healthy plant tissue. It is possible that some of the known phytoalexins may be able to move through the same routes taken by systemic fungicides, but this has not been investigated. What is not possible at present is to give any evidence that the known examples of cross-protection are caused by the movement of phytoalexins. The remote cross-protection in cucumber discussed in the previous section seems most unlikely to be mediated by phytoalexins because of the failure to detect such compounds in the Cucurbitaceae, as described in Chapter 5.

There are at present stronger indications of other types of translocated change, involving the diffusion of unknown substances from cells affected by the protecting organism to nearby cells, thus causing a change in sensitivity of these cells to normally virulent (compatible) fungi. This was the conclusion of Skipp & Deverall (1973) arising from observations and experiments on cross-protection in the anthracnose disease of bean. Following the observation that the protection was caused after the entrance of the avirulent and virulent races into the plant, as mentioned earlier in this chapter, it was found that the avirulent fungus would also exert its effect when it was added one day after the virulent fungus. By this means, it was possible to see that the virulent hyphae grew into the host cells before the avirulent germ-tubes caused nearby cells to undergo hyper-sensitivity. Once the latter had happened, normally compatible cells containing virulent hyphae also underwent hypersensitivity. This occurrence, even at a distance of several cells from the avirulent appressorium, suggested a change in the sensitivity of cells to normally compatible hyphae. Further evidence for a change in the sensitivity of bean cells around those which had undergone hypersensitivity to avirulent germ-tubes came from the use of heat treatments. Four days after inoculation of hypocotyls with the avirulent race, when penetrated cells were necrotic, the hypocotyls were given a heat shock of 50 °C for 30 seconds. As a result, all neighbouring cells became necrotic whereas those at a distance from the penetrated cells remained healthy. Thus the physiology of bean cells around hypersensitive cells had changed and they had become abnormally sensitive to either the presence of compatible hyphae or to heat. Similar observations of changed sensitivity of tobacco cells to heat around local lesions caused by a strain of TMV were made by Ross & Israel (1970).

The report by King, Hampton & Diachun (1964) that infection of leaves of the red clover *Trifolium pratense* by bean yellow mosaic virus affected their susceptibility to the powdery mildew *Erysiphe polygoni* is also relevant. Virus infection altered metabolism of the leaves so that they underwent hypersensitivity to the normally compatible powdery mildew. The visible responses of the host cells to the virus were the same as those reported by Smith (1938) for varieties of red clover genetically resistant to *E. polygoni*.

Thus there are a number of reasons why a factor, emanating from hypersensitive cells and able to change the sensitivity of susceptible cells so that they become resistant, should be sought. Some of the results of experiments described in the last chapter, concerning specific cross-protection factors in bean and an RNA from rust fungi in hypersensitive wheat leaves, may aid in this search.

The apparent implication of the remote cross-protection in cucumber is that two separate processes require investigation, at least in this phenomenon. One concerns the nature and action of a factor which passes from the inoculated leaf. The second concerns the nature of the change induced in the upper foliage. Is the acquired resistance associated with hypersensitivity of the

39

cells to attempted infection, or is some other type of change in host physiology involved?

PREVENTION OF HYPERSENSITIVITY BY HEAT-KILLED BACTERIAL CELLS

There remains the need to consider the questions posed by the series of discoveries following the demonstration by Lozano & Sequeira (1970) that heat-killed cells of an avirulent race of *Pseudomonas solanacearum* prevented hypersensitivity of tobacco leaves to normal cells of the same race. The areas of tobacco leaf injected with a minimum population of 3.5×10^7 cells/ml of an avirulent race normally become completely necrotic 24 hours later. This necrosis was prevented when the same areas were infiltrated with an equal concentration of heat-killed cells 18 hours before injection with live cells. Lower concentrations of heat-killed cells than live cells, and shorter intervals between infiltration and injection, were much less effective. Not only did the heat-killed cells prevent the necrosis of the leaf, but they also caused a more rapid decline in populations of the avirulent race than normally occurs as hypersensitivity develops. It was also noted that the preventive effect of the heat-killed cells was light dependent. Substantial prevention of necrosis was also caused by heat-killed cells of other races of *P. solanacearum*, and of *P. lachrymans* and *Xanthomonas axonopodis*, but not of *Escherichia coli*. These findings, along with further, more interesting discoveries by Lozano and Sequeira, should be considered with that of Loebenstein & Lovrekovich (1966) that heat-killed cells of *Pseudomonas syringae* interfered with local lesion formation by TMV in tobacco. Firstly, longer periods of time between infiltration with heat-killed cells and injection permitted the protective effect to spread not only into neighbouring areas of the treated leaf but later (within two days) to leaves immediately above this leaf. The preventive effect in the more remote leaves was seen by their response to injection as giving small necrotic spots rather than total necrotic collapse. Secondly, the heat-killed cells also prevented symptoms caused when virulent cells were injected into the treated area of leaf 24 hours later. Furthermore, the populations of these virulent cells declined instead of increasing rapidly as in untreated leaves.

Attempts have been made to separate the factor from heat-killed bacterial cells which prevents the development of hyper-sensitivity. Sequeira, Aist & Ainslie (1972) obtained several active crude extracts from the cells. A crude extract obtained by brief sonication of cells was completely effective, at a protein content of 0.1 mg/ml, when introduced into leaves seven hours before injection with the avirulent race. High molecular weight compounds separated from this extract by ethanol precipitation and fractionation on a column of Sephadex G-200 were active at a protein content of 2.4 mg/ml, even after heating at 95 °C for 10 minutes. The likely proteinaceous nature of the factor is shown by the destruction of activity by proteolytic enzymes, but first attempts to characterize and isolate the active protein were unsuccessful (Wacek & Sequeira, 1973).

This research, being mainly concerned with prevention of hypersensitivity, differs in its major theme from most of the phenomena described in this chapter. However, it is appropriate that it should be considered here because it is relevant to the debate concerning direct interference of one organism, albeit a dead one, with a pathogen, and it adds to the reports of systemically induced changes in response to infection in tobacco. Direct interference by blocking special sites in intercellular spaces is a possible mode of action when the need to infiltrate the same number of dead cells as living cells is considered, but this explanation is rendered improbable because of the delay of 18 hours before these dead cells can fully affect the reaction of the leaf. An induced metabolic change in the host thus seems the likely explanation of the phenomenon, and this is supported by the changes in response which occur in leaves remote from the site of treatment.

CONCLUSION

Although most of the earliest subjects for research in the study of cross-protection have not been re-investigated by modern methods, there are now clearly numerous convincing reports of localized protection against many important pathogens, and intriguing new claims of systemic protection in a few diseases. The major challenges for the future concern their exploitation in new methods of disease control and comprehension of their mode of action. In most cases, protection seems to be brought about by a change in host physiology and this give rise to the anticipation that similar changes might be induced by chemical applications. Deeper understanding of the natures of these changes in host physiology can only be advantageous to the intelligent development of new control procedures based on chemotherapy or biological control.

CHAPTER 5

Phytoalexins and their Induced Formation and Biosynthesis

The idea that plants produce protective chemical substances after infection was expressed by a number of research workers in the first half of this century, but the concept was formalized by Müller & Börger (1941). As explained in Chapter 4, their observations on hypersensitivity in potato cultivars resistant to the potato blight fungus, and their experiments on the induction of resistance in susceptible cultivars, led them to postulate the existence of phytoalexins. The term phytoalexin meant a warding-off compound produced by the plant, and phytoalexins were thought to form in hypersensitive potato tissue and to prevent further growth of the infection hyphae, but it was also considered that they might be of general occurrence in infected plants.

The first demonstration of the detection of a chemical entity as a phytoalexin was done by Müller (1958) working with the hypersensitive response of bean tissue to the soft-fruit pathogen *Monilinia fructicola*. Droplets of spore suspension were placed in the cavities of opened bean pods from which the seeds had been removed. The spores were observed to germinate and to cause death of some underlying cells within 24 hours. Infection droplets were collected after different intervals and were tested for their effects on new spores. The droplets became increasingly antifungal after incubation in seed cavities for 14 hours and completely fungistatic after 24 hours. The substance responsible for the antifungal activity could be extracted from combined infection droplets by partition with petroleum spirit, but it was not chemically characterized at the time.

The first characterization of a compound as a phytoalexin followed work with pea pods and *M. fructicola* which revealed a similarly extractable entity in infection droplets (Cruickshank & Perrin, 1960) to that found in bean pods. The compound from

43 4-2

the pea was isolated, crystallized, characterized as a pterocarpan and named pisatin (Perrin & Bottomley, 1962). Soon after this, the work with bean was re-examined and a closely related pterocarpanoid compound was isolated, characterized and named phaseollin (Cruickshank & Perrin, 1963a; Perrin, 1964). The discovery of these compounds was reviewed by Cruickshank (1963) who pointed out that concurrent work on other infected plants had similar implications. The sesquiterpenoid compound ipomeamarone from infected roots of the sweet potato *Ipomoea batatas* (Hiura, 1943; Kubota & Matsuura, 1953), the compound orchinol (Hardegger, Biland & Corrodi, 1963) from infected tubers of the orchid *Orchis militaris* and methoxy-mellein from infected roots of the carrot *Daucus carota* (Condon & Kuć, 1960, 1962) were therefore all regarded as phytoalexins. Thus, several different chemical compounds were soon recognized as phytoalexins, and all were relatively small molecules.

The involvement, in a much simpler way, of antifungal compounds in disease resistance had been suggested many years earlier in work with onions and the cause of smudge disease, *Colletotrichum circinans*. Antifungal compounds were found to diffuse into infection droplets on the outer surfaces of bulbs of the onion *Allium cepa* resistant to this fungus. The compounds were identified as protocatechuic acid (Link, Dickson & Walker, 1929; Angell, Walker & Link, 1930) and catechol (Link & Walker, 1933) and they were found to be present in the dead outer scales before infection. An important feature of this work was that progeny of crosses between resistant and susceptible cultivars of onion were tested for presence of these compounds. Association of the compounds with resistance in all progeny supported the hypothesis that they were the resistance factors. Similar hypotheses that particular pre-formed compounds are the bases of resistance in other plants have rarely been tested as extensively as in this early study.

Irrespective of the amount of evidence for a role of pre-formed factors in the resistance of some species and cultivars, it is philosophically difficult to conceive a role for such factors in the highly selective interactions between physiologic races of parasites and differential host cultivars, or in induced resistance. As discussed in Chapter 1, infection type is usually determined by a genetically based interaction between races and cultivars. Although pre-formed factors may be important in nature and of

44

value as a basis of some forms of resistance in new cultivars, they are of less fundamental interest as resistance factors than phytoalexins or other substances and processes which appear after infection.

Yet another way in which antifungal compounds might be involved in resistance is by their release from inactive precursors under the influence of enzymes normally separated from these precursors in different parts of healthy tissues. Thus the cereals, rye, wheat and maize *Zea mays*, contain glucosides of dihydroxy-benzoxazolinone and glucosidases, which interact in homogenized tissue resulting in the rapid release of antifungal aglucones (Hietala & Wahlroos, 1956; Virtanen & Hietala, 1959; Wahlroos & Virtanen, 1959). Those infections of cereal cells which cause mingling of cellular constituents in hypersensitive reactions might be expected to cause release of the active compounds. Comparable processes involving different types of compound are known to be possible in a number of other plant species, as reviewed by Ingham (1972, 1973). Release of antifungal compounds by these processes is embraced within the original concepts of phytoalexins as expressed by Müller & Börger (1941) and by Müller (1958). However, as far as is understood the compounds which were first recognized as phytoalexins are synthesized from remote precursors in hypersensitive tissues and because of this they command greater interest.

The main reasons for believing that compounds such as pisatin and phaseollin are synthesized and not released from immediate precursors are the slowness with which they accumulate and the absence of knowledge of any simple precursors in pea and bean. Although all antifungal compounds in plants may be of importance in defence mechanisms against infections, greatest interest surrounds phytoalexins which are likely to be synthesized by pathways activated after infection. Many ways can be conceived in which host and parasite metabolic systems may interact both in the activation of these pathways and in the regulation of different enzymatic steps within the pathways. Studies on phytoalexins are therefore stimulated by the thought that one is concerned not only with the agents of defence of plants but also with systems through which specific interactions of host and parasite might be mediated. Thus the remainder of this chapter will be concerned with knowledge of phytoalexins in those families in which they have been revealed and with some of the general problems posed concerning the processes and cellular sites for their formation. The next chapter will review understanding of their role in resistance and ways in which they might be involved in regulating specific interactions between plants and parasites.

The recognition of pisatin and phaseollin as phytoalexins in pea and bean led first to the investigation of several other legumes for phytoalexin production and then to a re-investigation of the original species for additional phytoalexins. Both lines of research have been highly productive. Pterocarpanoid phytoalexins were found in alfalfa *Medicago sativa* and the soybean *Glycine max*, the alfalfa compound being named medicarpin (Smith, McInnes, Higgins & Millar, 1971) and the soybean compound being first characterized as closely related to phaseollin and named hydroxyphaseollin (Sims, Keen & Honwad, 1972). This structure and name were then found to be incorrect and a revised structure of the soybean phytoalexin has now been proposed (Burden & Bailey, 1975). Further investigation of bean resulted in the characterization of three chemically related phytoalexins in addition to phaseollin. These additional bean phytoalexins were phaseollidin (Perrin, Whittle & Batterham, 1972), phaseollinisoflavan (Burden, Bailey & Dawson, 1972) and kievitone (Burden *et al.*, 1972; Smith *et al.*, 1973). In concurrent work, red clover was shown to produce two phytoalexins (Higgins & Smith, 1972), one being the same compound, medicarpin, as produced by alfalfa and the other, maackiain, which has also been detected as a minor product of infected pea in addition to pisatin (Stoessl, 1972). Other concurrent research has revealed a second isoflavonoid phytoalexin in alfalfa, namely sativan (Ingham & Millar, 1973) which has also been found in *Lotus corniculatus* together with a related compound, vestitol, by Bonde, Millar & Ingham (1973). Medicarpin has also been detected as a phytoalexin in *Canavalia ensiformis* (Keen, 1972).

Thus many legumes produce at least one and usually several phytoalexins as listed in Table 7, which is arranged according to the tribal classification of legumes (Hutchinson, 1964). It will be seen that a number of legume species produce the same com-

TABLE 7. *Phytoalexins in some tribes of the Leguminosae*

Tribe*	Species	Phytoalexin	Reference
34. Diocleae	*Canavalia ensiformis*	Medicarpin	Keen (1972)
37. Phaseoleae	*Phaseolus vulgaris*	Phaseollin	Perrin (1964)
		Phaseollidin	Perrin *et al.* (1972)
		Phaseollinisoflavan	Burden *et al.* (1972)
		Kievitone	
	P. lunatus	Phaseollin	Cruickshank & Perrin (1971)
	P. radiatus		
	P. leucanthus		
	Vigna sinensis	Phaseollidin	Bailey (1973)
		Kievitone	
		Phaseollin	
38. Glycineae	*Glycine max*	A soybean pterocarpan	Burden & Bailey (1975)
40. Vicieae	*Cicer arietinum*	Medicarpin	Keen (1975a)
		Maackiain	
	Vicia faba	Wyerone acid	Letcher *et al.* (1970)
		Wyerone	Fawcett *et al.* (1971)
	Pisum sativum	Pisatin	Perrin & Bottomley (1962)
		Maackiain	Stoessl (1972)
	P. arvense	Pisatin	Cruickshank & Perrin (1965)
	P. elatius		
	P. abyssinicum		
	P. fulvum		
42. Trifolieae	*Medicago sativa*	Medicarpin	Smith *et al.* (1971)
		Sativan	Ingham & Millar (1973)
	Trifolium pratense	Maackiain	Higgins & Smith (1972)
		Medicarpin	
	T. repens	Medicarpin	Cruickshank *et al.* (1974)
	Lotus corniculatus	Sativan	Bonde *et al.* (1973)
43. Loteae		Vestitol	

* According to the classification of Hutchinson (1964).

47

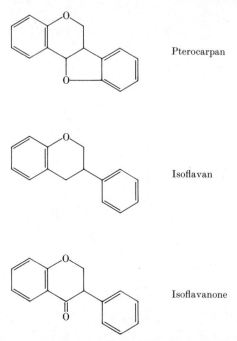

Pterocarpan

Isoflavan

Isoflavanone

Fig. 1. Basic structures of pterocarpans, isoflavans and isoflavanones.

pound, and this is most marked with respect to medicarpin. Comprehension of this is aided by consideration of the structures and chemical affinities of these phytoalexins as reviewed by Van Etten & Pueppke (1976). The basic structures of pterocarpans, isoflavans and isoflavanones are presented in Fig. 1, and the structures of some of the known phytoalexins arranged in these respective classes in Figs. 2, 3 and 4. A range of side substitutions on the basic ring structures occur, but it will be seen that medicarpin is one of the simplest of the pterocarpans in this regard.

Table 7 also shows that the species *Vicia faba* produced two phytoalexins, named wyerone and wyerone acid, which have not yet been discussed. Wyerone was first characterized as an antifungal component of extracts of broad bean seedlings which had been germinated for eight days between wet sacks. However, Fawcett, Firn & Spencer (1971) failed to find wyerone in healthy broad bean leaves but showed that it accumulated in substantial amounts after infection by *Botrytis fabae.* The closely

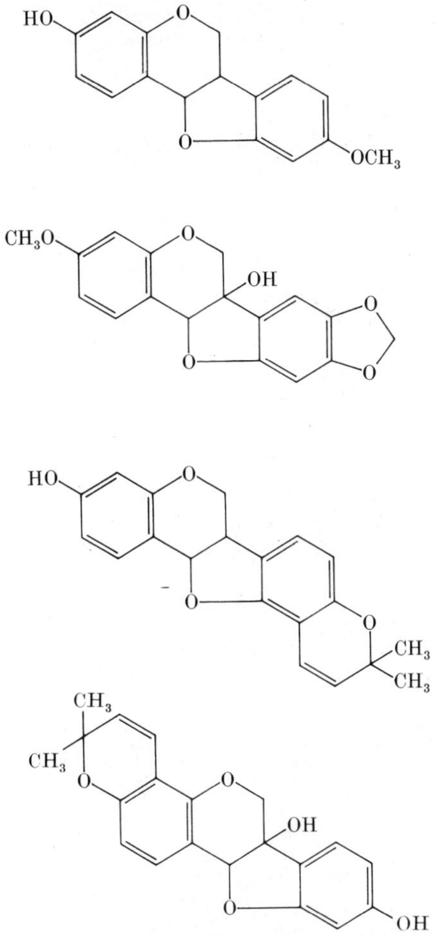

Medicarpin

Pisatin

Phaseollin

Soybean phytoalexin
(glyceollin, see p. 73)

Fig. 2. Structures of some pterocarpanoid phytoalexins.

related wyerone acid had earlier been detected as a phytoalexin in infected broad bean (Letcher, Widdowson, Deverall & Mansfield, 1970). It is now clear that both compounds must be regarded as phytoalexins because they form in antifungal quantities after infection of broad bean tissues by *Botrytis* spp. (Mansfield, Hargreaves & Boyle, 1974) and are absent in healthy tissues. The structures of the compounds shown in Fig. 5 are quite different from, and present a strange phytochemical contrast with, the isoflavonoid phytoalexins of the other members of

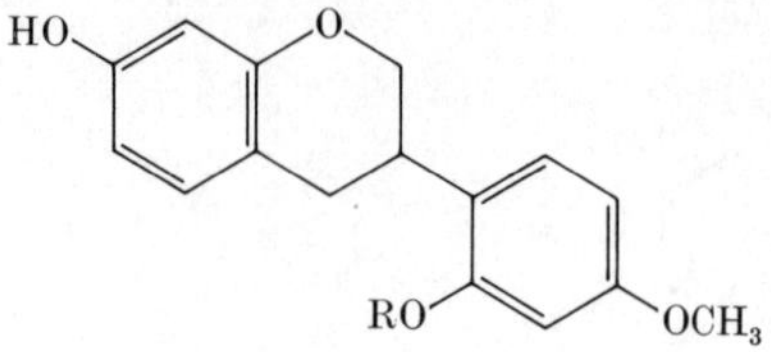

where R = H,
Vestitol

where R = CH$_3$
Sativan

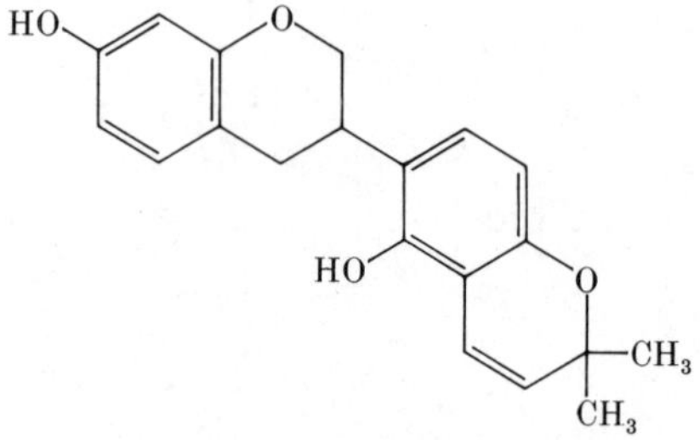

Phaseollinisoflavan

Fig. 3. Structures of some isoflavans as phytoalexins.

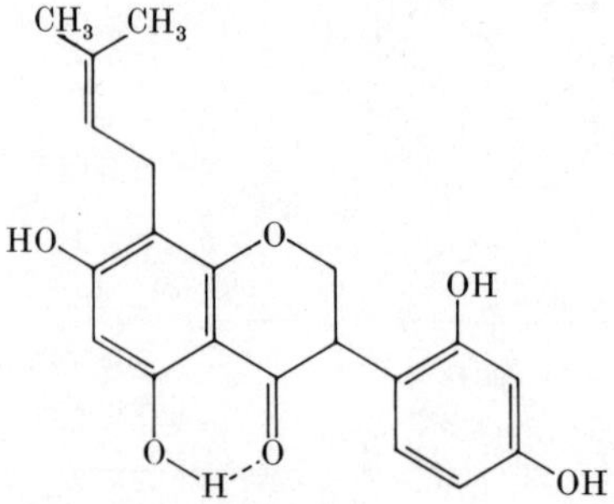

Kievitone

Fig. 4. Structure of the isoflavanone kievitone, a phytoalexin from bean.

the Leguminosae studied to date. Wyerone is the methyl ester of the acetylenic keto furanoid fatty acid. Their structures indicate that different types of phytoalexin have evolved within the legumes.

The detection of more than one phytoalexin in many legume species is now common. It is possible that fungal metabolism might be directly responsible for the conversion of one compound to others in some infected plants, as discussed more fully in the next chapter. However, research by Bailey & Burden (1973) and by Bailey (1973) indicates that bean and cowpea

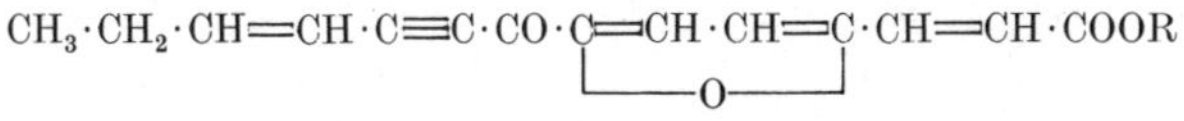

$$CH_3 \cdot CH_2 \cdot CH \!=\! CH \cdot C \!\equiv\! C \cdot CO \cdot \underset{\underset{O}{\rule{2.2em}{0.5pt}}}{C} \!=\! CH \cdot CH \!=\! C \cdot CH \!=\! CH \cdot COOR$$

where R = H,   Wyerone acid

where R = CH_3 Wyerone

Fig. 5. Structures of the acetylenic keto furanoid aliphatic phytoalexins from broad bean *Vicia faba*.

produce four and three phytoalexins respectively following infection by tobacco necrosis virus, which is assumed not to have a metabolic capacity itself. Therefore in these species at least, it is clear that several phytoalexins can be formed by the plant tissues after the virus has induced some necrosis. The significance of the often observed association between necrosis and phytoalexin formation will be discussed again later.

PHYTOALEXINS OF THE SOLANACEAE

It will be recalled that phytoalexins were envisaged by Müller & Börger (1941) to form in potato cells expressing hypersensitivity to avirulent races of *Phytophthora infestans*. However, no chemical entity was demonstrated as a phytoalexin in potato until the work of Tomiyama *et al.* (1968) with the same disease, and this led to the characterization of the terpenoid phytoalexin, rishitin (Katsui *et al.*, 1968), which was also found in infected tomato tissue (Sato, Tomiyama, Katsui & Masamune, 1968). Later, two further terpenoid phytoalexins were detected in potato, namely lubimin (Metlitskii, Ozeretskovskaya, Vul'fson & Chalova, 1971) and phytuberin (Varns, Kuć & Williams, 1971; Coxon, Curtis, Price & Howard, 1974). Numerous other terpenoid compounds also accumulate in resistant potato tubers after infection, presumably as part of a general stimulation of terpenoid synthesis during hypersensitivity (Kuć, 1976). Pepper produces the terpenoid phytoalexin, capsidiol (Gordon, Stoessl & Stothers, 1973) in response to fungal infection, and the same compound is formed by leaves of two species of tobacco *Nicotiana tabacum* and *N. clevelandii*, following infection with tobacco necrosis virus (Bailey, Burden & Vincent, 1975). Another related compound, glutinosone, has been isolated as an antifungal product of leaves of another tobacco species, *N. glutinosa*, bearing small necrotic lesions caused by TMV (Burden, Bailey & Vincent, 1975).

51

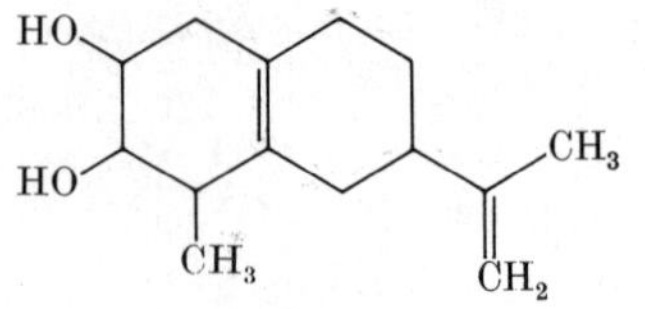

Rishitin

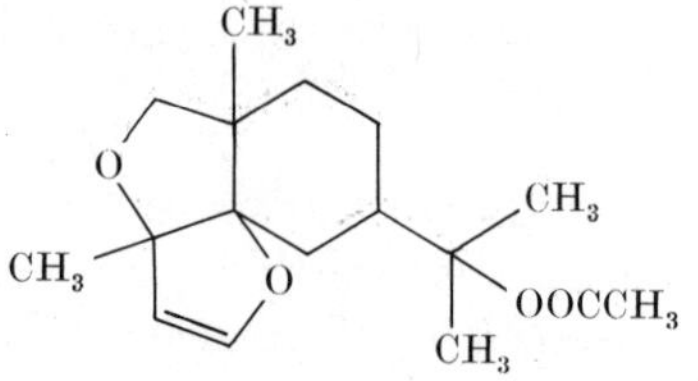

Phytuberin

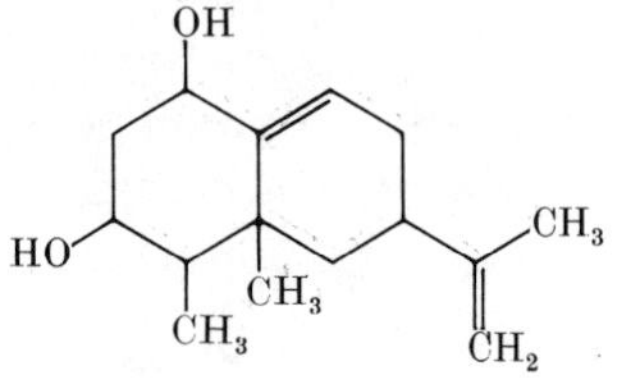

Capsidiol

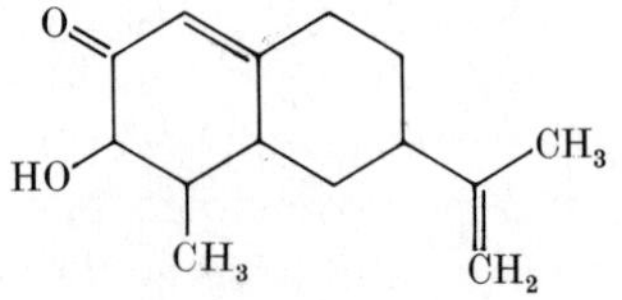

Glutinosone

Fig. 6. Terpenoid phytoalexins of the Solanaceae.

Thus, although fewer species have been examined in the Solanaceae, similar phenomena to those found in the Leguminosae have been revealed. Related terpenoids (shown in Fig. 6), as distinct from isoflavonoids, occur in different species. Potato and toabcco, which have been studied most intensively, produce several active compounds as does bean. Again, there seems to be a close association between necrosis and phytoalexin accumulation, and virus-induced necrosis in tobacco is again effective in causing phytoalexin formation.

PHYTOALEXINS IN OTHER FAMILIES

Only a limited number of other families in the plant kingdom, and few species within them, have been studied for phytoalexin production, and these are listed in Table 8. Substituted

TABLE 8. *Phytoalexins in plant families other than the Leguminosae and Solanaceae*

Family	Species	Phytoalexin	Reference
Chenopodiaceae	*Beta vulgaris*	2',5-Dimethoxy-6,7-methylenedioxyflavanone	Geigert *et al.* (1973)
Malvaceae	*Gossypium barbadense*	2'-Hydroxy-5-methoxy-6,7-methylenedioxyisoflavone Vergosin Hemigossypol	Zaki, Keen, Sims & Erwin (1972)
Umbelliferae	*Daucus carota*	3-Methyl-6-methoxy-8-hydroxy-3,4-dihydroisocoumarin	Condon & Kuć (1962)
	Pastinaca sativa	Xanthotoxin	Johnson, Brannon & Kuć (1973)
Convolvulaceae	*Ipomoea batatas*	Ipomeamarone	Kubota & Matsuura (1953)
Compositae	*Carthamus tinctorius*	Safynol Dehydrosafynol	Allen & Thomas (1971*a*) Allen & Thomas (1971*b*)
Orchidaceae	*Orchis militaris*	Orchinol	Hardegger *et al.* (1963)
	Loroglossum hircinum	Hircinol	Gäumann (1964)

53

isoflavones occur in one member of the Chenopodiaceae, and substituted isocoumarins in the Umbelliferae. The cotton species *Gossypium barbadense* in the Malvaceae produces at least two naphthaldehydes in response to infection. Two poly-acetylenic aliphatic alcohols may be regarded as phytoalexins in the safflower *Carthamus tinctorius* although one of these compounds, safynol, can be obtained in low yield in healthy tissues (Allen & Thomas, 1971*a*). Acetylenic compounds occur in many plants and can possess antifungal properties (Sorensen, 1961; Bu'lock, 1964). Further research may reveal the importance of polyacetylenic compounds as agents of defence in the Compositae and other families. Orchid tubers were the subject of some of the most significant early work on induced resistance, as discussed in Chapter 4, and this work eventually led to the characterization of phenanthrene phytoalexins, orchinol and hircinol, in two orchid species, *Orchis militaris* and *Loroglossum hircinum*, respectively. In resolving some earlier disputes about the source of these compounds, Gäumann & Jaag (1945) and Gäumann & Kern (1959*a, b*) showed that there was a negligible amount of antifungal substances in healthy tubers but that high concentrations of the phytoalexins formed after infection.

So few plants outside of the Leguminosae and the Solanaceae have been reported on for phytoalexin formation, that it is impossible to know how widely the process occurs in higher plants. This lack of information probably reflects the fact that other families have not been studied, but some investigations may have yielded negative results which have not been described. In this context it is useful to record a personal failure to find any post-infectional increase in amounts of some uncharacterized antifungal substances detected in healthy fruits and leaves of *Cucumis sativus* and *Cucurbita pepo* in the Cucurbitaceae, even after hypersensitive responses to *Colletotrichum lagenarium*. This indication that phytoalexin formation can not be detected in the Cucurbitaceae using methods which readily reveal phytoalexins in the Leguminosae should be considered alongside other personal experiences with wheat in the Gramineae. In this case, the use of a number of extraction procedures coupled with separatory and bioassay methods successful in work on phytoalexins in the Leguminosae, revealed an antifungal compound in wheat seedlings which had expressed resistance to stem rust *Puccinia graminis* f.sp. *tritici*. However,

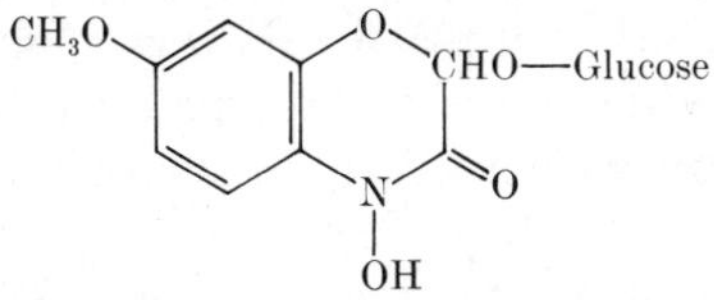

Fig. 7. Structure of the glucoside of dihydroxymethoxybenzoxazolinone (DIMBOA) in wheat *Triticum aestivum*.

the properties of the compound detected suggest that it is related to the benzoxazolinones discussed earlier and, therefore, that it was released from a glucosidic precursor in hypersensitive cells (Fig. 7). Thus, it is unlikely that wheat can synthesize a phytoalexin in the way that legumes and solanaceous plants seem to do. The implication of these two examples for the general question under consideration in this paragraph is that phytoalexin formation is not a universal phenomenon in the plant kingdom. Just as it has been shown that quite different types of molecule have evolved as phytoalexins, even within the Leguminosae, it would seem that several different mechanisms of defence may have evolved among the higher plants. However, much more research is needed on this question because the numbers of reports of failure to detect phytoalexins are so few compared with the numbers of successful investigations.

INDUCTION AND SITES OF PHYTOALEXIN FORMATION

In the discussion of the phytoalexins in the Leguminosae and the Solanaceae, the close association between virus-induced necrosis and the production of phytoalexins was emphasized. This raises questions concerning the means by which phytoalexin formation is induced following different types of infection, and the cellular sites at which phytoalexins are formed. Is necrosis the cause of phytoalexin formation, or is it an associated phenomenon? Do phytoalexins form in dying or dead cells, or in living cells? These questions are not only important for a complete understanding of the natural sequence of events in infected plants, but also have a bearing upon the problems involved in raising the concentrations of protective chemicals in plants when epidemics of disease threaten.

Soon after pisatin was discovered, it was shown by Cruickshank & Perrin (1963*b*) that the ions of some heavy metals induced the formation of pisatin. This was followed by demonstrations that a wide range of organic molecules achieved the same thing (Perrin & Cruickshank, 1965; Bailey, 1969; Hadwiger & Schwochau, 1970). Adequate explanations of the biochemical means by which the formation of pisatin is induced are not available, although proposals about effects on nuclear DNA have been put forward (Hadwiger, Jafri, von Broembsen & Eddy, 1974). Of more immediate relevance to an understanding of the sequence of events during parasitism was the detection of a metabolite of the fungus *Monilinia fructicola*, which was active at low molarities in inducing formation of phaseollin by seed cavities of bean pods (Cruickshank & Perrin, 1968). This fungus, a soft-fruit pathogen, was that used in the original experiments whereby phytoalexin formation was first demonstrated as bean pods underwent hypersensitivity. The implication of the work with the fungal metabolite, named monilicolin A, was that this substance might be responsible for causing phaseollin formation during infection. Greater interest in monilicolin A results from the demonstration that it will cause phaseollin to form without first killing cells in the bean pod (Paxton, Goodchild & Cruickshank, 1974). Thus the hypothesis may be advanced that a substance of parasitic origin acts to induce the formation of phytoalexins by live host cells. Such a hypothesis is also consistent with the concept, discussed briefly in Chapter 1, that a product of the gene for avirulence activates a defence mechanism in plants by interacting with a product of a host gene for resistance. The defence mechanism might be phytoalexin formation by live cells.

Some caution is needed in ascribing a role in the host–parasite interaction for a molecule capable of inducing phytoalexin formation. Because so many organic molecules have this ability, as indicated above, it is not surprising that active inducers can be obtained from fungal mycelium or culture filtrates. It is necessary to consider whether the inducer might act in the same way during the process of infection. A more compelling reason for believing that a naturally acting inducer had been isolated would be the demonstration that it possessed the same specificity as the parasite from which it was obtained. Thus, a specific inducer would be active only on host cultivars resistant to the avirulent race which produced the inducer.

The demonstrations that fungal metabolites induce

phytoalexin formation by healthy cells must be contrasted with the observations that phytoalexins began to accumulate in a number of infected plants at about the time that necrosis was first noticed. These observations are based upon studies of the infection of *Vicia* by *Botrytis* (Mansfield & Deverall, 1974*b*), of *Phaseolus* by *Colletotrichum lindemuthianum* (Bailey & Deverall, 1971; Rahe, 1973) and by *Uromyces appendiculatus* (Bailey & Ingham, 1971) and of several legume and *Nicotiana* spp. by viruses, as mentioned earlier. Thus it seems possible that phytoalexin formation is induced as a consequence of cellular death in these host–parasite relationships.

Phytoalexin formation has been caused in two types of experiment in which physical injury has been used to simulate the damage caused by infection. Thus, moderate bruising of broad bean leaves caused necrosis and phytoalexin formation within 24 hours (Deverall & Vessey, 1969), although severe bruising was ineffective perhaps because cells were killed too rapidly. Certain forms of freezing injury also caused phaseollin to form in bean (Rahe & Arnold, 1975), although other types of injury such as pin-pricking and scratching were ineffective stimuli for pisatin formation in pea pods (Cruickshank & Perrin, 1963*b*). Thus, phytoalexin formation could be brought about in at least two ways during pathogenesis: by substances diffusing from or present on walls of infecting hyphae and by processes associated with the death of host cells in response to the activity of hyphae.

The fact that phytoalexin formation is often associated with cellular necrosis during pathogenesis has caused consideration of the idea that phytoalexins may be produced by dead or dying cells. Elaboration of new molecules in such cells seems unlikely, although hydrolyses and oxidations might be expected to occur. Indeed, it has been suggested by Rathmell & Bendall (1972) that final interconversions of isoflavonoids, as envisaged in phaseollin production, might be catalysed by peroxidases likely to be active in dying cells. Living cells around dying cells are the most likely sites of more elaborate syntheses, and it has been noted by a number of workers that the cytoplasm of these cells has unusual properties, such as heightened sensitivity to heat and to normally compatible hyphae (Ross & Israel, 1970; Skipp & Deverall, 1973) and increased densities (Mercer, Wood & Greenwood, 1974). The detection of an unusual fluorescence in the vacuoles of live broad bean cells adjacent to necrotic cells infected by *Botrytis cinerea*, has great relevance to the problem of sites of synthesis of phytoalexins (Mansfield *et al.*, 1974). The emission spectrum of this fluorescence was identical to that of pure solutions of wyerone and wyerone acid. It seemed likely to Mansfield and his colleagues that the phytoalexin had been made in these live cells and had not diffused from the necrotic cells because there was not a ring of fluorescent cells around the dead cells.

There are now several reasons for concluding that phyto-alexins are made by living cells. The major unresolved question concerns the stimulus for their formation. Clearly phytoalexin formation can be brought about in several ways. Virus infections and physical damage must activate a control system in the plant which induces phytoalexin production. Fungi may secrete inducing molecules, perhaps highly selective in their action on host cultivars, which trigger the synthesis of phytoalexins. How closely linked is the process of host cell death to the formation of phytoalexins? Several different relationships seem possible, such as

    (i)   death of a host cell may cause synthesis of phytoalexins in neighbouring cells;

    (ii) molecules diffusing from a fungus may induce both phytoalexin formation and cell death, perhaps at increasing doses, respectively;

    (iii) phytoalexin accumulation may cause death of host cells as speculated by Mansfield *et al.* (1974).

THE BIOSYNTHESIS OF PHYTOALEXINS

As explained at the beginning of this chapter, there is reason to believe that phytoalexins are synthesized in the plant by metabolic pathways which are activated following infection. Although much information is available concerning the metabolic pathways by which isoflavonoid and terpenoid com-pounds are produced in some organisms, little work has been done on the activity of these pathways in diseased plants, and no work has been done on the biosynthesis of the acetylenic com-pounds after infection. As a result of this, we lack the evidence necessary, firstly, to corroborate the notion of the biosynthesis of phytoalexins from remote precursors as distinct from conver-sion or hydrolytic release from close precursors and, secondly, to evaluate the role of particular enzymatic steps in these pathways. New methods of disease control based upon manipulation of naturally-occurring systems might become possible in the light of understanding of the means of induction of phytoalexin formation and of appreciation of any key processes in phytoalexin biosynthesis. This section will briefly describe the pathways likely to be important in biosynthesis, and the nature of the evidence for the involvement of these pathways after infection.

Existing understanding of isoflavonoid biosynthesis in plants has been well summarized by Van Etten & Pueppke (1976). Grisebach (1965) demonstrated that the basic flavonoid skeleton was derived from the products of two metabolic pathways, the acetate–malonate and the shikimic acid routes. Phenylalanine is produced via shikimic acid and is then deaminated by the enzyme phenylalanine–ammonia lyase to cinnamic acid. Cinnamic acid is then prepared for condensation with acetate units to form either a chalcone or its isomeric flavanone (Grisebach & Hahlbrook, 1974). Subsequent steps in the formation of isoflavans and pterocarpans have not been demonstrated unequivocally, but a possible metabolic grid involving these compounds was presented by Van Etten & Pueppke (1976) based on a scheme proposed by Wong (1970). Evidence for the involvement of these steps in diseased plants is very limited. Several investigators have shown some transfer of isotopic ^{14}C from phenylalanine to the isoflavonoids pisatin and phaseollin (Hadwiger, 1967; Hess, Hadwiger & Schwochau, 1971). There is abundant evidence for the greatly enhanced activity of the enzyme, phenylalanine–ammonia lyase in some infected plants, and Rathmell (1973) has shown that this occurs before phaseollin accumulates in bean infected by *Colletotrichum lindemuthianum*. The involvement of the enzyme chalcone–flavone isomerase could not be established in phaseollin biosynthesis, but greatly increased peroxidase activity is a very common accompaniment of infection of many plants and this may play a role in the formation of isoflavonoids from immediate precursors (Rathwell & Bendall, 1972). Thus, although an enhanced rate of production of cinnamic acid seems to occur in infected legumes, the likely subsequent steps from cinnamic acid to pterocarpans and isoflavans have scarcely been examined.

An alternative source of flavonoids in legumes has been

suggested to be glycosidic precursors by Olah & Sherwood (1973) who found that glycosidases, probably of fungal origin, increased greatly in activity in alfalfa leaves following infection by *Ascochyta*. However, Olah & Sherwood (1971) found that the flavonoid glycosides did not decrease in concentration in diseased leaves as might be anticipated if hydrolysis was occurring. Further examination of metabolism in diseased alfalfa leaves seems desirable to ascertain the origin of medicarpin and sativan as phytoalexins and to evaluate the possible enhanced synthesis (which it seems necessary to postulate in order to explain the observed effects) of flavonoid glycosides after infection. From a related type of investigation, Rathmell (1973) was forced to conclude that the infection of bean specifically stimulated isoflavonoid synthesis, because he was unable to detect any concurrent changes in flavonoids released from glycosides.

Terpenoid biosynthesis is considered to arise from acetate and mevalonate in plants, and evidence has been obtained of a major movement of isotopic ^{14}C from these precursors into rishitin and related compounds in hypersensitively reacting potato tuber discs (Shih & Kuć, 1973). The interesting concept that the compatibility of the infecting organism determines whether the post-infectional metabolism of infected potato tissue will be switched towards terpenoid accumulation in resistant tissue, or steroglycoalkaloid production in susceptible tissue, has been proposed by Shih, Kuć & Williams (1973) and this will be discussed further in the next chapter. The biosynthetic pathway from acetate and mevalonate to ipomeamarone in infected sweet potato tubers has been outlined, and the role of dehydro-ipomeamarone as a precursor of the phytoalexin has been partially studied (Oguni & Uritani, 1974). Thus, more substantial evidence is available for the activation of terpenoid biosynthesis in infected plants, although it has not yet proved possible to identify the key step in this activation.

CONCLUSION

Abundant evidence for the existence in some plant families of phytoalexins with a diversity of chemical structures has been obtained. Evidence for their role in controlling fungal development in certain host–parasite relationships will be discussed in the next chapter. Some important questions remain to be answered concerning the means by which the formation of phytoalexins is induced after infection, and confirmation of the biosynthesis of phytoalexins from remote precursors is needed. Understanding of these processes may permit insight into new ways of controlling plant diseases by manipulation of naturally-occurring processes. However, it can not be concluded that phytoalexins are universal in the plant kingdom, and it may be that several different means of chemical defence have evolved in the higher plants.

CHAPTER 6

Role of Phytoalexins in Defence Mechanisms

The fact of the existence of phytoalexins in at least some plant families, as discussed in the previous chapter, does not prove their role in the defence mechanisms of plants. It is tempting to ascribe a function for biologically active molecules isolated from living organisms, but important questions can be raised about the involvement of such molecules in the complex processes within living tissues. The general nature of these questions will be presented first, before moving on to a discussion of the nature of the evidence for the involvement of phytoalexins in disease resistance and in the processes controlling specialization in parasitism.

The need for caution in asserting a role in disease resistance for each anti-microbial compound extracted from a plant was emphasized by Daly (1972). Thus it is desirable to know whether the compound occurs at those micro-sites within a plant where it might contact a parasitic bacterium or fungus. The places in plants where different parasitic micro-organisms cease growth in resistant tissues were described in Chapter 3, but little information is available about the location of phytoalexins in the intercellular spaces, cell walls, cytoplasm, organelles and vacuoles of living cells, or about their movement in and around dead cells. Many phytoalexins are lipophilic and it is legitimate to ask whether they occur at these important micro-sites in plant tissues not only in sufficient concentration but also in the appropriate form to affect growth of parasites. It is also necessary to know whether phytoalexins accumulate at the appropriate stage during infection to cause the observed cessation of growth of the parasite. The possibility of being misled about the sequence of events during infection and expression of resistance was suggested by the demonstration that the growth of several parasitic fungi in plant tissues can be stopped by the application of an antibiotic and that, as an apparent consequence, the plant cells die and then phytoalexins accumulate (Király *et al.*, 1972). Under this interpretation, phytoalexins would be a consequence and not a cause of the expression of resistance. Thus many legitimate doubts may be held about the role of phytoalexins in natural defence processes, and there is a need for ingenious experimentation to test the hypothesis that phytoalexins cause parasites to stop growing in resistant plants.

RELATIONSHIPS BETWEEN PHYTOALEXIN ACCUMULATION AND RESISTANCE

Phytoalexins and diseases of bean

The phytoalexins of bean have been extensively examined for a possible role in resistance to avirulent races of the anthracnose fungus *Colletotrichum lindemuthianum*. Phaseollin accumulates in high concentrations after cells have reacted hypersensitively to fungal penetration from appressoria (Bailey & Deverall, 1971; Rahe, 1973). The phaseollin is restricted to the small areas of necrotic tissue at infected sites, as revealed by extractions of excised sections of stems, but its location inside necrotic cells and/or their immediate neighbours has not been established. If the phaseollin is in the necrotic cells alone, its concentration six days after inoculation is more than 3000 μg/ml, which greatly exceeds the 10 μg/ml which prevents germ-tube growth *in vitro* (Bailey & Deverall, 1971). However, it is possible that the phaseollin forms in surrounding live cells and never enters the necrotic cells, as discussed in Chapter 5. Phaseollin accumulates in the two to three day-period after the first symptoms of cell death appear and this coincides with the period when germ-tubes of the fungus have been seen to slow in growth rate and to become restricted inside the necrotic cells (Skipp & Deverall, 1972). Phaseollin is not the only phytoalexin to accumulate in hypersensitive tissue, and Bailey (1974) has shown that three other active compounds form in substantial amounts at the same time. Thus phytoalexin formation is a distinctive feature of the development of a low infection type and is associated with the early host necrosis, which is the first obvious indication of a high degree of incompatibility between fungus and host.

Although there is much evidence to suggest that resistance is expressed through the action of the bean phytoalexins on germ-tubes, it remains possible that they do not contact the germ-tubes at all, or do so too late to cause cessation of growth. Investigations on the localization of the phytoalexins within cells in the infection court are desirable, although as yet, no methods of detecting the bean compounds by histochemistry or by micro-spectrophotometry have been developed. One way of assessing the involvement of phytoalexins as factors which stop fungal growth in hypersensitive cells, is by the use of phytoalexin-tolerant variants or mutants of an avirulent race of the fungus. Large numbers of wild-type spores, or those which have been exposed to a mutagen, could be screened for their ability to grow on culture medium in the presence of the bean phyto-alexins. Phytoalexin-tolerant forms would then be cultured separately and tested for their ability to infect resistant bean cultivars. Interest would centre on those forms which had retained the ability to cause the hypersensitive response and phytoalexin formation. If phytoalexins are normally responsible for the cessation of growth in the necrotic cells, the tolerant forms should not be limited to these cells but should grow on through the tissues. This experiment has not yet been done, but it should prove a direct test of the hypothesis that phytoalexins are responsible for the expression of resistance to the anthrac-nose fungus in bean.

By contrast with the development of the low infection type, the hyphae of virulent races of *Colletotrichum* are compatible with susceptible host cells for at least several days during which time substantial intracellular growth is made without any adverse effects on the protoplasts (Skipp & Deverall, 1972). No phaseollin accumulates throughout this phase of parasitism in the development of a high infection type (Bailey & Deverall, 1971; Rahe, 1973). An analysis of the factors responsible for this major difference in phaseollin accumulation in the two infec-tions types is needed to aid understanding of the regulation of phytoalexin formation in infected plants. As discussed in Chap-ter 5, it will be necessary to examine the apparent association between necrosis and phytoalexin formation, and to consider the possible production by avirulence genes of incompatibility factors which interact with products of resistance genes in caus-ing either necrosis or phytoalexin synthesis as a general host response to a specific recognition.

Any hypothesis concerning the molecular interactions which

64

determine infection type, must take into account the facts that necrotic limited lesions can develop, and that fungal growth can become restricted in susceptible hypocotyls after a period of compatibility between host cells and intracellular hyphae. The continuation of fungal growth depends upon some unknown internal conditions and upon environmental factors, as revealed in two separate studies. In dark-grown hypocotyls, the fungus continued to spread at some infection sites, but not at others, where brown limited lesions appeared. Rahe (1973) found that phytoalexins accumulated rapidly where fungal growth was limited, but not where the fungus continued to spread. In light-grown hypocotyls, Bailey (1974) found that symptom development was temperature-dependent. At 17 °C, the fungus continued to spread and caused the entire hypocotyl to rot, and very little phytoalexin could then be found. At 25 °C, fungal growth was restricted to the places where early mycelial growth had been established, and these sites formed sunken brown lesions containing high concentrations of phaseollin and phase-ollinisoflavan. This evidence strongly suggests that phytoalexin accumulation plays a role in lesion spread by preventing further fungal growth. An answer to the question of what factors control the formation of phytoalexins at some infection sites, but not others, can not be provided at present, but once again the association of necrosis and phytoalexin accumulation must be noted. Rahe (1973) observed that phaseollin could not be detected until 17 to 18 hours after the appearance of necrosis, which suggests that the stimulus for its formation under these circumstances is a product of the host–parasite interaction, and perhaps of dead cells. Furthermore, he noted that the amounts of phaseollin accumulating were more closely related to the surface areas of the lesions than to any other dimension. This might suggest that the stimulus from the dead cells causes phaseollin formation in live cells around the edge of the lesion. The effect of higher temperatures might be to accelerate the death of host cells which are colonized by hyphae, and thereby to cause phytoalexin formation. Another possibility, that the fungus is able to metabolize phytoalexins at lower temperatures, has been considered by Bailey (1974) and there is evidence that *Colletotrichum lindemuthianum* can metabolize phaseollin (Burden, Bailey & Vincent, 1974), as discussed later.

It may be useful to view the common susceptibility of bean

seedlings to fungal diseases in cold soils as a possible indication of the general failure of defence systems to act in the bean at lower temperatures. Among the fungi responsible for seedling diseases of the bean are *Rhizoctonia solani* and *Fusarium solani* f.sp. *phaseoli*. Although an analysis of the effects of temperature on phytoalexin metabolism has not been reported for these diseases, it is known that phytoalexins accumulate when limited lesions form. Kievitone begins to appear at 30 °C in young lesions caused by *Rhizoctonia solani* and then accumulates in high concentrations, along with phaseollin, as the lesions mature and become brown and limited (Smith, Van Etten & Bateman, 1975). An interesting difference in similar lesions caused by *F. solani* f.sp. *phaseoli* is the absence of kievitone, although phaseollin again accumulates (Van Etten & Smith, 1975). Thus, in limited lesions caused by the three different pathogens of bean, phaseollin occurs in all but is accompanied by phaseollinisoflavan in *Colletotrichum* infections and kievitone in *Rhizoctonia* infections. The reasons for and implications of these different accumulations of phytoalexins are not understood at present. However, it seems very likely that phytoalexin accumulation creates an antifungal environment in limited lesions, and that this, possibly acting with other factors, prevents further fungal growth in susceptible bean hypocotyls at higher temperatures.

Phytoalexins and Phytophthora *resistance in soybean*

A number of studies have been performed on changes in a soybean phytoalexin following infection of hypocotyls from different cultivars with races 1 and 2 of *Phytophthora megasperma* var. *sojae*. Mycelial inocula were placed in small wounds at central points in the hypocotyls, and two days later, high infection types were characterized by extensive rotting of the hypocotyls, whereas low infection types were characterized by local red-brown lesions around the points of inoculation. A feature of the development of both infection types was the detection of phytoalexin about 10 to 12 hours after inoculation (Keen, 1971). Accumulation continued for another 36 hours, although at much faster rates and in much higher concentrations in the wound areas where limited lesions were forming. This accumulation was similar to that observed in beans where hypersensitivity or limited lesion formation occurred, and similarly, implies that it is probably involved in the restriction of growth of the fungus.

A marked difference from the bean research was the observation of early formation of phytoalexin in high infection types. Clearly, unlike bean cells infected with compatible races of *Colletotrichum*, soybean cells begin to produce phytoalexin when infected by compatible races of *Phytophthora* before accumulation becomes suppressed by an unknown process. Whether this difference is caused by the inoculation procedure or by the nature of the host–parasite interaction has not been revealed.

An interesting development in soybean research has resulted from an investigation of fungal products which cause phytoalexin to form when applied to hypocotyls. Culture fluids of each race are quite active in this respect (Keen, 1975*b*) but there is preliminary evidence that the products of avirulent races are particularly effective when applied to resistant hypocotyls. Gel filtration of the products of an avirulent race yielded a fraction which was much more active in causing phytoalexin formation in a resistant than a susceptible cultivar. This claim is based, at present, on slender evidence thus considerable caution must be exercised in accepting that the hypothetical specific inducer or elicitor of phytoalexin formation has been found.

Phytoalexins and Phytophthora *resistance in potato*

Another host–parasite relationship, involving specific interactions between particular cultivars and different physiologic races, which has been extensively studied for the role of phytoalexins, is that between the potato and the blight fungus *Phytophthora infestans*. In early experiments, Tomiyama *et al.* (1968) found that inoculation of cut slices of potato tubers with an incompatible race caused hypersensitivity and rishitin formation, whereas infection by a compatible race caused only a trace of rishitin to form in the same period when the hyphae were growing in cells without inducing visible reactions (Kitazawa & Tomiyama, 1969). As in bean research, the location of rishitin in or around hypersensitive cells has not been revealed but, if it is at highest concentrations in necrotic cells, it would create strongly antifungal conditions. Again, as in the studies of *Colletotrichum lindemuthianum* in hypersensitive bean cells, germtubes of *Phytophthora infestans* continue to grow, albeit increasingly slowly, for some hours after potato cells have died. Thus the sequence of host cell death and cessation of fungal growth in both these host–parasite interactions is contrary to that implied by the work of Kiraly *et al.* (1972) and is consistent with the hypothesis that rishitin is involved in expression of resistance in potato. These events have been confirmed by the work of Varns, Kuć & Williams (1971) and Metlitskii *et al.* (1971), who revealed that phytuberin and lubimin are also produced as phytoalexins in hypersensitive tuber tissue. Kuć (1976) has stated that the expression of resistance not only involves the production of these antifungal terpenoid compounds but also many other terpenoids.

Some interesting ideas have been developed concerning the way in which virulent hyphae affect susceptible potato cells so that the phytoalexins do not form. Firstly, it must be appreciated that phytoalexin formation in potato tubers is a common response to many treatments (Varns, Currier & Kuć, 1971). These treatments include inoculation with fungi other than *Phytophthora infestans* and application of extracts of the mycelium of both virulent and avirulent races of *P. infestans.* These extracts cause phytoalexin formation in tubers of many potato cultivars, including those which lack any of the so-called *R* genes for resistance to potato blight. However, virulent hyphae do not cause phytoalexin formation when growing in susceptible cells (Varns *et al.*, 1971). Furthermore, prior inoculation with a virulent race of *P. infestans* diminished the amount of phytoalexin which formed when tubers were inoculated with an avirulent race (Varns & Kuć, 1971). It was suggested, therefore, that virulent hyphae suppress phytoalexin formation in susceptible tubers, and divert metabolic pathways so that non-antifungal compounds are produced (Kuć, 1976).

THE SIGNIFICANCE OF PHYTOALEXIN METABOLISM BY FUNGI IN DISEASE DEVELOPMENT

The involvement of phytoalexins in hypersensitivity and lesion limitation has already been discussed, but it is also possible to consider their role in diseases, in which the first symptom is a local lesion from which successful forms of the parasite are eventually able to spread throughout the infected plant. Perhaps the simplest disease of this type is leaf spot of the broad bean, caused by *Botrytis* spp. An early consequence of penetration of epidermal cells by germ-tubes of the fungus is necrosis, resulting in the appearance of a small brown lesion. The fungus *B. cinerea* is confined to these small lesions, but virulent isolates of *B. fabae* spread by growing in dead cells, killing surrounding cells and then growing in these (Mansfield & Deverall, 1974a). The phytoalexins, wyerone acid and wyerone, accumulate rapidly in and around the necrotic cells of the epidermis during the first two days after inoculation with *B. cinerea*, and the use of micro-spectrophotometry has revealed the production of these phytoalexins by live cells adjacent to necrotic cells at this time (Mansfield *et al.*, 1974). Mansfield and his colleagues suggested that wyerone acid might also be phytotoxic and that it might play a role in causing the death of host cells, as well as preventing development of fungal hyphae. The frequently expressed idea, that the phytoalexins form and limit germ-tube growth in the first cells penetrated and killed, is very plausible but requires testing further. Some wyerone acid-tolerant mutants of *B. cinerea* should be able to spread from these initial infection sites if wyerone acid is the natural principle limiting their progress, but this experiment has not yet been attempted. Support for the idea that wyerone acid is a limiting factor to fungal growth, comes from the demonstration that its concentration declines rapidly as *B. fabae* begins to spread from the infection sites (Mansfield & Deverall, 1974b). The decline occurs first at the sites of inoculation while some wyerone acid is accumulating around the lesions, and then continues in the peripheral tissue as the fungus advances. Wyerone acid is replaced in the lesion, and then throughout the leaf, by a reduced form of wyerone acid, which is less antifungal than the phytoalexin. The chemical reductions occur in the acetylenic and keto-groups of the wyerone acid molecule (Mansfield & Widdowson, 1973) and are almost certainly caused by the fungus in the leaf, because the same process is carried out by the fungus when fed wyerone acid in a culture tube. *B. cinerea*, on the other hand, appears to be unable to carry out the reduction of wyerone acid. Thus this study highlights the probable role of wyerone acid in the limita-tion of *B. cinerea* and indicates a metabolic capacity which might be an essential contributor to the pathogenicity of *B. fabae*. The use, in pathogenicity tests, of naturally-occurring variants or induced mutants of *B. fabae* that lack this capacity to reduce wyerone acid, should permit this hypothesis to be tested in a direct way.

A further interesting aspect of this work concerns the way in which pollen grains and their exudates promote the pathogenicity of *Botrytis cinerea*, so that it is able to colonize broad bean leaves in the same way as *B. fabae* (Chou & Preece, 1968). Wyerone acid accumulates in high concentrations in leaves colonized by *B. cinerea* in the presence of pollen, but its antifungal action seems to be prevented by exudates of the pollen (Mansfield & Deverall, 1971). A possible mode of action of these exudates is prevention of uptake of wyerone acid by the fungus (Deverall & Rogers, 1972) but this is a problem requiring further study. Although not completely understood, this dramatic effect, possibly significant in the field if anthesis coincides with other conditions favouring the disease, suggests further that phytoalexin formation plays a key role in the normal defence of the broad bean plant against *Botrytis* spp.

The study of the pathogenicity of *B. fabae* provides a strong indication of the importance of fungal metabolism of phytoalexins in plants. Other studies have revealed metabolic capacities in pathogenic fungi which have similar implications for understanding mechanisms of pathogenicity.

One of these concerns the metabolism of the pterocarpanoid medicarpin, and it started with an investigation of the differences between several leaf-spot pathogens of alfalfa leaves. *Stemphylium botryosum* and *S. loti* enter these leaves via stomata and produce substomatal vesicles from which intercellular secondary hyphae arise and cause necrosis in contacted host cells. Hyphae of these fungi develop in and around these necrotic cells to greater and lesser extents respectively (Pierre & Millar, 1965; Higgins & Millar, 1968). By way of complete contrast, *Helminthosporium turcicum* and *Colletotrichum phomoides* failed to develop further than the necrotic epidermal cells over which they had produced appressoria. It was as a result of an analysis of the antifungal nature of the infection droplets containing *H. turcicum* on alfalfa leaves, that medicarpin was isolated and characterized as an alfalfa phytoalexin (Smith *et al.*, 1971). Earlier experiments had shown that germinated spores of *S. botryosum*, but not of *H. turcicum*, had the ability to cause the disappearance of the then uncharacterized phytoalexin *in vitro* (Higgins & Millar, 1969a). Subsequent investigations by Higgins (1972) revealed that medicarpin was produced by alfalfa leaves in response to infection by each leaf-spot pathogen, but that no medicarpin accumulated in infection droplets on the leaf surface containing *S. botryosum*. Medicarpin is probably metabolized by *S. botryosum* in alfalfa leaves, because the mycelium is insensitive to the phytoalexin in culture and can convert it to vestitol (Higgins & Millar, 1969*b*; Steiner & Millar, 1974). However, as pointed out by Higgins (1972), the sites of formation and suspected degradation of medicarpin in and around infected cells have not been investigated because of the lack of techniques capable of monitoring these events at micro-sites in leaves bearing small lesions.

The metabolic detoxification of phaseollin probably occurs while *Fusarium solani* f.sp. *phaseoli* infects bean hypocotyls because a product, hydroxyphaseollone, is readily detected (Van Etten & Smith, 1975) as it is when phaseollin is fed to the fungus in culture (Heuvel & Van Etten, 1973; Heuvel *et al.*, 1974).

High levels of antifungal benzoic acid can be found in arrested lesions caused by *Nectria galligena* in unripe apples, but the concentrations decline as the fungus begins to spread in the ripe fruit (Brown & Swinburne, 1971). Benzoic acid is not metabolized by apple tissues but is converted by the fungus to *p*-hydroxybenzoic acid and protocatechuic acid. The less antifungal *p*-hydroxybenzoic acid accumulated in rotted tissue, and protocatechuic acid, which is even less active, could be detected in advanced rots (Brown & Swinburne, 1973). It is suggested by these authors that, as acidity declines and sugar content increases in ripening fruit, the fungus is able to metabolize benzoic acid and thus to grow as the fungitoxicity is decreased.

The phytoalexin capsidiol is rapidly oxidized to the much less active capsenone by isolates of *Fusarium oxysporum*, both in culture and in pepper fruit, and this process may be an important step in the colonization of this parasite in peppers (Stoessl, Unwin & Ward, 1973).

By contrast, the ability of the fungus *Aphanomyces euteiches* to cause rapidly spreading lesions in pea tissues can not be explained at present in terms of its ability to tolerate or to metabolize the phytoalexin pisatin. Pueppke & Van Etten (1974) have shown that the fungus causes the formation of high concentrations of pisatin, greatly in excess of those which prevent growth *in vitro*, yet continues to grow. A decline in the pisatin level was recorded at later stages of lesion development, but it was not of an order which could explain the apparent anomaly of a pisatin-sensitive fungus growing readily in a pisatin-rich environment.

Studies on the metabolism of the pterocarpanoid phytoalexins

71

in vitro have often preceded demonstrations that they are metabolized by the pathogens in plants. Thus three different metabolic fates of phaseollin have been revealed. *Stemphylium botryosum* converts phaseollin to the analogous isoflavan phaseollinisoflavan (Heath & Higgins, 1973; Higgins, Stoessl & Heath, 1974). *Colletotrichum lindemuthianum* hydroxylates phaseollin to hydroxyphaseollin (Burden, Bailey & Vincent, 1974) whereas *Fusarium solani* f.sp. *phaseoli* carries out a different hydroxylation to hydroxyphaseollone (Heuvel *et al.*, 1974). It appears likely that one enzyme in *Stemphylium botryosum* can carry out similar conversions of phaseollin, medicarpin and maackiain to the corresponding 2'-hydroxyisoflavan (Higgins, 1975), because the converting system for one of the compounds in spore suspensions could be induced by any of the others. However, the converting enzyme has not been isolated to test this likelihood. Isolations of the enzymes responsible for any conversions of phytoalexins have been unsuccessful to date, and the only unequivocal activity in a cell-free preparation of a fungus was that of *Leptosphaerulina briosiana* with respect to medicarpin (Higgins, 1972).

PHYTOALEXINS AND BACTERIAL DISEASES OF PLANTS

Phytoalexins were first conceived of as antifungal principles, but there is no reason why they should not be involved in the response of plants to bacterial infections. Indeed, some of the phenomena associated with resistance to some bacterial pathogens, as discussed in Chapters 3 and 4, would be explicable in terms of the formation of bacteriostatic substances during hypersensitive reactions. The earliest research on phytoalexins did not encourage this idea, because pisatin formation was not stimulated by bacteria in pea pods (Cruickshank & Perrin, 1963*b*), and pisatin had no effect on the growth of several species of phytopathogenic bacteria in culture (Cruickshank, 1962). However, research with other plants in recent years has stimulated further interest in the involvement of phytoalexins in bacterial infections.

This new research began with the demonstration that, during the hypersensitive response of bean leaves to *Pseudomonas phaseolicola*, phaseollin accumulated to concentrations of the order of 200 μg per g fresh weight of leaf (Stholasuta, Bailey,

Severin & Deverall, 1971). Phaseollin was not active against *P. phaseolicola* in liquid culture, but it was considered that some of the, as then, unknown compounds, formed in noticeable amounts in hypersensitive tissue might be antibacterial. This consideration was corroborated by the work of Lyon & Wood (1975), who revealed that extracts of leaves showing a similar response contained six fractions which were active against *Brevibacterium linens* when bioassayed directly on thin-layer chromatograms. Three of these fractions were then found to be active against *P. phaseolicola*. The activity of one of these fractions was attributed to coumestrol and was considered the most important, but the source of the activity in the other two fractions has not yet been identified. Although coumestrol also accumulated in leaves expressing high infection types at later stages of infection, its absence in healthy leaves and its marked accumulation in the first day of the development of low infection types indicate that its probable role is that of an antibacterial phytoalexin in bean.

Coumestrol has also been implicated in the hypersensitive response of soybean leaves to *Pseudomonas glycinea* (Keen & Kennedy, 1974). Not only coumestrol but also the related compounds daidzein, sojagol and the phytoalexin recently renamed glyceollin (Keen, personal communication) accumulated rapidly following infection. Glyceollin particularly, but also coumestrol, were inhibitory to the growth of the bacterial pathogen on agar media. This property and the fact that the concentrations of the two compounds increased greatly in leaves before bacterial multiplication became restricted, are the bases for regarding them as antibacterial phytoalexins.

The terpenoid phytoalexins, rishitin and phytuberin, are produced in potato tubers following inoculation with *Erwinia carotovora* (Lyon, 1972); and rishitin, but not phytuberin, was inhibitory to the growth of the bacterium in culture (Lyon & Bayliss, 1975). Rishitin was also found to be bactericidal against the organism suspended in peptone water, a medium used to maintain viability of bacteria without permitting an increase in cell numbers. Although rishitin may contribute to the resistance of potato tubers to bacterial soft-rot stored in the air, no correlation was found between the rishitin concentrations and the extents of resistance in different cultivars (Lyon, Lund, Bayliss & Wyatt, 1975). Further study of the concentration and localization of rishitin, in and around the advancing areas of the rot, is necessary to reveal the amounts which might contact the bacterium *in vivo*.

The terpenoid phytoalexin capsidiol has also been isolated from pepper fruit infected with *Erwinia carotovora* (Ward, Unwin & Stoessl, 1973), but in this case no indications of antibacterial activity could be seen by means of tests on agar and in liquid media.

CONCLUSION

The evidence that phytoalexin accumulation in numerous members of the Leguminosae and Solanaceae is responsible for the inhibition of fungal or bacterial growth in hypersensitive tissues, and in limited lesions, has been cautiously but favourably reviewed. Further work is required on the cellular localization of phytoalexins relative to the infecting organisms, and the hypothesis that the substances cause the expression of resistance might be tested by means of phytoalexin-tolerant variants or mutants. The factors involved in the accumulation of phytoalexins at early stages in the development of low infection types, but not high infection types, remain to be revealed. The rival ideas, that avirulent parasites specifically induce phytoalexin formation and the possibly linked host necrosis in resistant cells, or that virulent parasites specifically suppress phytoalexin formation and again the possibly linked rapid necrosis in susceptible cells, require experimental investigation.

Another basis for specialization by necrotrophic parasites is the evolution of an enzyme system for the degradation of phytoalexins which accumulate at an early stage of parasitism. Several examples of this type were discussed, and once again the hypotheses can be tested by using variants or mutants with or without the particular enzymatic capacity.

CHAPTER 7

Mediation of Host–Parasite Specificity

Parasitic fungi and bacteria require general attributes to fit themselves for the parasitic, rather than the saprophytic, habit, and possess special features which enable them to be virulent on particular taxonomic groupings of host plants. The general properties of parasites include the ability to enter a plant by a specialized route, the capacity to synthesize enzymes for breaching barriers presented by cuticles and cell walls, the ability to obtain nutrients from the host, possibly via haustoria, and perhaps, the capacity to tolerate or to metabolize anti-microbial compounds in the host plant. Fine degrees of specialization might sometimes depend upon success or failure of these processes, but it should be clear from Chapters 1, 2 and 3 that the fate of parasitism is usually dependent upon the matching of very specific information, determined by complementary genes in host and parasite and expressed after penetration into host cells. The major question for this chapter concerns the intermolecular means by which the single genes in host and parasite express themselves. In what form is the information based on the DNA in both partners compared?

It seems logical that expression of specificity is mediated by nucleic acids, peptides and proteins, although lipoproteins, polysaccharides, glycoproteins and glycolipids may also be able to act as mediation factors. The essential feature of these molecules is that they are made up of many subunits arranged in particular sequences and thus suited to recognize matching molecules in the other partner in the host–parasite relationship. Evidence is needed to show that molecules of this type have a special activity in the early stages of attempted parasitism.

Mediation of specificity might occur through a recognition reaction linked to other host reactions which then accept or reject the intruding parasite. For example, a recognition reaction might determine whether or not a host cell will react hypersensitively to a parasite. A rejection reaction might be death of the host cell and starvation of the parasite if a biotroph, or it might be based on synthesis of a phytoalexin or release of an anti-microbial compound from a precursor. Alternatively, specificity might be mediated through a reaction which recognizes and accepts or rejects in the same process.

In considering these alternative ideas, it should be borne in mind that the most specific genetic interactions are usually for the expression of a low infection type as depicted in Tables 2 and 3 (Chapter 1). The simplest hypothesis is that primary products of the genes for avirulence and resistance interact to cause incompatibility. Thus a positive interaction for recognition and rejection would seem most likely, except in those host–parasite relationships where virulence and susceptibility are the dominant characteristics.

The existing state of knowledge in host–parasite physiology prevents the presentation of a coherent account of the occurrence or nature of recognition factors. A few years ago, almost nothing useful could be written on the subject. With advances in techniques and concepts, a number of completely unrelated experiments with different host–parasite systems have now been done. Each bears upon the problem under consideration without permitting any complete synthesis of ideas concerning the nature of recognition. These experiments and their implications can be considered under a number of headings:

  (i)   host-specific toxins;
  (ii) common antigens;
  (iii) message-containing cross-protection factors;
  (iv) an RNA as recognition product of an avirulent rust;
  (v)   elicitation or suppression of phytoalexin formation.

HOST-SPECIFIC TOXINS

Consideration of the sequence of events during many host–parasite interactions and of the occurrence of hypersensitivity leads to the idea that an avirulent parasite secretes or bears a toxin which disrupts the membranes and cytoplasm of resistant cells but specifically not of susceptible cells. Recognition might be mediated by toxins acting on, or just in advance of, contact between haustoria and host protoplasts. However, it is not possible to present convincing evidence for the detection of any such toxins, but it must be written that few appropriate investigations have been done. Experiments should be performed to seek these toxins in germination fluids or in extracts of walls of avirulent parasites, and to reveal their presence using bioassays which are chosen to show early effects on host cells. Possible bioassay materials are tissue slices, dispersed cells in culture and naked protoplasts of resistant and susceptible plants.

A considerable body of evidence exists to show that some parasites produce host-specific toxins with exactly the opposite properties to those envisaged above. For example, the culture filtrates of a number of fungi, especially *Helminthosporium* spp. and *Periconia circinata*, yield toxins which specifically damage and kill susceptible host cells. It is useful to describe briefly the history of their discovery and to refer to their remarkable properties as recognition factors and apparently as determinants of virulence.

The first host-specific toxin was discovered as a result of the appearance of a new disease of oats, which only affected cultivars bred for resistance to crown rust caused by *Puccinia coronata* (Meehan & Murphy, 1946). Resistant cultivars carried a gene from the Victoria cultivar of oats, which conferred hypersensitivity to the rust. Thus, when these cultivars began to succumb in the field to a new isolate of *Helminthosporium*, the disease was called Victoria blight and the pathogen, *H. victoriae*. The fugus infects the base of the stem but causes symptoms on the shoot system, suggesting the action of a toxin. Culture filtrates were toxic to oat cultivars carrying the Victoria gene (Meehan & Murphy, 1947). A toxin was isolated by Pringle & Braun (1957) and shown to prevent growth in a bioassay of roots of susceptible oat seedlings at 0.0002 μg/ml (Scheffer & Pringle, 1963) but to have no effect on roots of resistant seedlings at 400000 times this concentration (Kuo, Yoder & Scheffer, 1970).

Almost overwhelming evidence that the highly active host-specific toxin must be regarded as a primary determinant of virulence was provided by Scheffer & Yoder (1972) in a review of the many investigations by Scheffer and his colleagues on this subject. The most persuasive argument is based on correlations between virulence and presence of toxin. All parasitic isolates which produced the toxin in culture were virulent, and there was a complete correlation between virulence and toxin production among a collection of wild types and mutants and among progeny of a cross between virulent and avirulent mutants. It is surprising that there were no exceptions, because it is likely that a toxin-producing mutant might grow in culture and yet fail to possess another essential attribute for parasitism apart from the ability to make toxin, as discussed in the first paragraph of this chapter. Further pertinent evidence is the restoration of virulence to avirulent spores by addition of toxin to inoculation sites, and again, complete success with all mutants is surprising. The detection of toxin in spores and their germination fluids shows that it is present at the appropriate stage of parasitism to have a primary influence. The very rapid action in causing release of electrolytes from susceptible tissue and in bursting susceptible protoplasts supports the notion that the toxin is a primary factor in establishing the parasite in the host. Unfortunately for the presentation of a complete account of the toxin, it has not yet been possible to characterize the toxin chemically because of its instability during extraction and purification (Pringle, 1972).

The search for other host-specific toxins has been so successful, that eight examples could be cited by Scheffer & Yoder (1972). Four of these toxins emanate from *Helminthosporium* spp. which cause diseases in members of the Gramineae. A fifth toxin has been obtained from *Periconia circinata*, also a pathogen of a plant in this family. All show marked specificity although none has the activity of the *H. victoriae* toxin.

Study of one of the toxins, discovered by Steiner & Byther (1971), a product of the sugar cane pathogen *Helminthosporium sacchari*, has resulted in some rapid advances concerning the detection of a specific receptor site in susceptible cultivars of the sugar cane *Saccharum officinarum*. Radioactive toxin was used to test for the abilities of different types of extract from susceptible and resistant sugar cane leaves to bind the toxin in an equilibrium dialyser (Strobel, 1973a). Binding ability was found in a preparation of membranes from susceptible leaves only. Treatment of these membranes with a detergent released a protein which also possessed the capacity to bind toxin. A comparable preparation of protein from resistant leaves was inactive (Strobel, 1973b). Evidence to support the hypothesis that the protein isolated from susceptible leaves was the toxin-receptor site came from two experiments. In both, various solutions were infiltrated for 24 hours into leaves through their lower cut ends before application of toxin to central positions on the leaves (Strobel & Hess, 1974). In one experiment, opposite halves of a leaf were fed either antiserum to the protein or control serum. No symptoms developed on the half fed antiserum, implying that the receptor site had been protected from the toxin. In the second experiment, the leaf tissue was flushed with the same type of detergent as that used to dislodge receptor protein from extracted membranes. Not only did detergent prevent the leaves from expressing symptoms but also the fluids recovered from the upper cut ends of the leaf contained the receptor protein. Persuasive evidence has thus been put forward that a specific receptor site for a host-specific toxin has been detected and isolated. Some of the properties of the receptor protein have been studied and compared with those of an apparently slightly different protein in resistant leaves (Strobel, 1973a, b).

The implications of the work with host-specific toxins are that susceptibility of a host depends upon the presence of a specific receptor site and resistance depends on its absence. Resistance must therefore be thought of as a passive feature in these cases, but some important questions remain with respect to the roles of the *Helminthosporium victoriae* and *H. sacchari* toxins. Yoder (1972) has discussed reasons for believing that, in the early stages of infection by *H. victoriae*, reception of the toxin by the susceptible host conditions living cells to provide a favourable environment for the fungus, whereas non-reception by resistant cells leaves an unfavourable environment in which the fungus cannot grow beyond penetrated cell walls. However, the essential features of this environment which make it conducive or otherwise for growth have not been defined. Is the fungus stimulated in one or inhibited in the other? Is susceptibility therefore induced or is resistance suppressed by the action of the toxin? The questions about the toxin of *H. sacchari* are even more basic and they must firstly be concerned with its role in gaining access for the fungus to susceptible tissues. Is it the key to the establishment of infection or is it merely the cause of the striking secondary symptoms, the orange-red runner lesions, which appear above points of infection or injection of toxin?

The knowledge of the inheritance of virulence and susceptibility as dominant traits in the relationship between *Helminthosporium victoriae* and oats permits the advance of the hypothesis that the toxin is the recognition factor produced by the gene for virulence, and the presumed receptor site in the host is the product of the gene for susceptibility. For no other host–parasite relationship can a better substantiated hypothesis be proposed for the mediation of specificity.

Reverting now to the type of toxin anticipated at the beginning of this section, the only evidence for substances which affect resistant, but not susceptible, cells concerns high molecular weight products of *Cladosporium fulvum*. This fungus exists as a number of physiologic races specialized to attack particular tomato cultivars. High molecular weight compounds were partially purified from culture filtrates of the different races. These compounds were infiltrated into discs cut from tomato leaves previously fed with radioactive phosphate, and the rate of leakage of isotope from the infiltrated discs was measured. Compounds from avirulent races were specifically able to cause rapid loss of isotope from resistant leaf discs. Products of virulent races had no effect, relative to controls, on leakage from susceptible discs. These experiments imply that the fungus produces toxins specific to resistant host cells (van Dijkman & Kaars Sijpesteijn, 1973), but it must be stated that the evidence, for their involvement as factors in parasitism, is far less substantial compared with the evidence concerning host-specific toxins active on susceptible cells. Until much more experimental work has been completed, this demonstration must be regarded as intriguing rather than convincing, although it is the type of indication which is sought. Experimental work on secretions from, or components of, germ-tubes of some of the many other fungi which cause hypersensitivity in resistant cultivars would be most welcome. Preliminary investigations of *Colletotrichum lindemuthianum* (Skipp & Deverall, 1973) and *Phytophthora infestans* (Sato *et al.*, 1968; Varns & Kuć, 1971; Kiraly *et al.*, 1972) have revealed complete non-specificity of the culture filtrates and mycelial extracts in their toxicity to bean and potato cultivars respectively. More refined methods for the extraction and assay of cultivar-specific toxic products are clearly necessary to investigate their existence in most fungi, and to confirm the claims for *Cladosporium fulvum*.

The concept that parasites and their hosts share common antigens has arisen in the study of parasites of both animals and plants. For example, Dineen (1963) revealed some antigenic relatedness between the sheep and its parasitic worm *Haemonchus contortus*. This type of relatedness may have evolved in the parasite to protect it from the immunological system in the animal which would recognize and destroy non-self. It is more surprising that the concept has arisen in plant pathology because there is no known comparable immunological process in plants. Furthermore, the notion of compatible host and parasite sharing a similar type of molecule is at odds with the simplest hypothesis based on the best-known genetic interactions that the most specific interaction is for incompatibility.

In plant pathology, research on sharing of common antigens was started by Doubly, Flor & Clagget (1960) following the proposal by Flor (1956) of the concept of the gene-for-gene relationship between flax cultivars and physiologic races of the flax rust *Melampsora lini*. Nearly isogenic lines for resistance to several races of flax rust were produced by repeated back crossing of resistant cultivars into a susceptible cultivar. Globulin antigens were prepared from each isogenic line and from each race of rust, and antisera were produced to each batch of antigens. Precipitin tests were run between all combinations of antigens and antisera. The results could be interpreted to show that some antigens were shared by susceptible cultivars and the races virulent on those cultivars.

A number of other host–parasite combinations have been examined for sharing of common antigens. Only one investigation has failed to reveal common antigens in host and parasite, and this concerns alfalfa and *Corynebacterium insidiosum* (Carroll, Lukejic & Levine, 1972). Among some remarkable positive findings was the discovery that *Xanthomonas malvacearum* had more in common antigenically with its host, cotton, than it did with other *Xanthomonas* spp. (Schnathorst & DeVay, 1963). Serological relatedness is usually considered an indicator of taxonomic affinity so this surprising report gives impetus to the view that the shared antigens of plant and parasite have functional roles in parasitism. Also impressive is the finding that not only does the maize smut *Ustilago maydis* share antigens with its host, maize, but also with young oat seedlings which it was found able to parasitize (Wimalajeewa & DeVay, 1971). A more remote antigenic relationship was found with older oat tissue which was not parasitized.

The study of *Ustilago maydis* and maize (Wimalajeewa & DeVay, 1971) is the only one in which both the location and nature of the common antigens were considered. The antigens were traced to the ribosomes, and they were deduced to be protein and not RNA. If the common antigens in ribosomes have functional significance in specificity and virulence, early interactions between ribosomes of host and parasite must be envisaged. This location appears to preclude simple ideas of common antigens functioning as recognition factors on surfaces of contacting host and parasite.

Thus correlations have been established between the serological properties of host and parasite in all but one of the combinations so far studied. The existence of the correlations does not prove that antigens play an essential role in parasitism and compatibility, and it is possible that the common antigens are either artefacts of the experiments or consequences of parasitism, with no regulatory role. Where bacterial parasites have been studied, there is a possibility that supposedly healthy plants carry a low population of virulent bacteria. Similarly, fungal spores and host cultivars might carry a common contaminant bacterium, which would show as the common antigen. Although such artefacts could arise in experiments, they seem unlikely to contribute the common antigens revealed in the experiments with different flax cultivars and rust races. Alternatively a sensitive serological technique might detect a component of the host plant, nuclear or cytoplasmic, acquired and carried by a parasite as a consequence of parasitism.

The essential and causal roles of common antigens in host–parasite compatibility have not been established. Roles for common antigens may be envisaged as the compatibility factors necessary before an association can occur between living cells of a plant and a parasite, as discussed by DeVay, Charudattan & Wimalajeewa (1972). It seems desirable now to advance, from the phase of research which has been concerned with correlations, to attempts to assess the significance and role of the antigens. One line of progress might be through a genetic approach, in which a parasite is bred for virulence in culture and then examined for a constant association between a particular antigen and virulence. A biochemically based approach would be the isolation of the common antigens by combined use of techniques for protein separation and serology, followed by measurement of effects of the antigens on resistant and susceptible tissues, cells or protoplasts.

SPECIFIC CROSS-PROTECTION FACTORS

The previous two sections have concerned detection of factors present in parasite and host before contact. This section is concerned with the detection of early products of the host–parasite interaction, which may be involved in passage of specific messages to neighbouring cells. These products may be responsible for the induction of the process of resistance in the host plant.

Specific factors affecting cross-protection were revealed in experiments by Berard, Kuć & Williams (1972) using infection droplets containing spores of *Colletotrichum lindemuthianum* on the surfaces of bean hypocotyls. Droplets containing spores of an avirulent race were incubated for 60 hours, collected, sterilized by micro-filtration, concentrated and termed 'diffusates'. Diffusates were placed on surfaces of new hypocotyls of the same cultivar and then overspotted with spore suspensions of a virulent race 18 hours later. As a consequence, no lesions developed. The diffusates were shown not to be directly antifungal to the virulent race, which germinated and penetrated epidermal cells. The diffusates changed the usual course of the host–parasite interaction so that growth of the virulent race was stopped in the penetrated, normally susceptible, cells. Diffusates from infection drops containing virulent spores were ineffective. The specificity of the factor in the diffusates from incompatible interactions was shown by its effectiveness against virulent races only on the same cultivar from which it was derived.

A further refinement was introduced into this work by the use of a larger range of cultivars. It was then found that the protection factor was also effective on different cultivars provided that they possessed the same genes for resistance as the donor cultivar. A cultivar bearing genes for resistance to two different races was shown to produce two protection factors

83

which were distinguished by their effects on cultivars carrying the genes separately (Berard, Kuć & Williams, 1973).

The specific cross-protection factors are uncharacterized and seem likely to remain so while their source is a small volume of combined infection droplets. Useful progress on their nature may be achieved with the aid of enzymes specific for the destruction of different types of message-bearing macro-molecule. The type of change that induces normally susceptible cells to become resistant is also unknown. From observations made during the experiments, hypersensitivity did not occur in the cells with induced resistance. This contrasts with the observations, described in Chapter 4, that premature necrosis of bean cells accompanied the protection caused by avirulent germ-tubes (Skipp & Deverall, 1973). As discussed in Chapters 5 and 6, phytoalexin formation follows necrosis of infected bean cells and is likely to cause cessation of hyphal growth in hypersensitive cells. However, further research is needed to implicate phytoalexin formation in the restriction of hyphae in infected cells which are not necrotic, although it will be recalled that a fungal peptide, monilicolin A, can cause phytoalexin formation in apparently healthy bean cells, and broad bean phytoalexins have been detected in live cells adjacent to necrotic infected cells. If phytoalexins are absent from the cells affected by the specific cross-protection factor, then the existence of another process capable of checking fungal growth will have been discerned.

Hypersensitivity is closely associated with normal resistance of bean cells to the anthracnose fungus as well as with induced resistance caused by avirulent races (Skipp & Deverall, 1972, 1973). The relevance, therefore, of the specific cross-protection factors and their mode of action (apparently in the absence of hypersensitivity), to the normal process of resistance is uncertain. However, the demonstration of their existence is intriguing and challenging. If further evidence on their nature and mode of action can be produced, this will have bearing on research on specificity, induced resistance and the controversy concerning the causal role of hypersensitivity in resistance.

Experiments have been conducted with wheat leaves undergoing resistance to stem rust to seek an early product of the host–parasite interaction which can affect the way in which normally susceptible cells react to infection. The experiments by Rohringer, Howes, Kim & Samborski (1974) used a wheat cultivar bearing the Sr_6 gene for resistance, and utilized the temperature sensitivity of the expression of this gene in order to obtain large numbers of hypersensitive cells. The wheat was inoculated with an avirulent race of rust and incubated at 26 °C so that the mycelium grew and pushed haustoria into many host cells. After three days the temperature was changed to 20 °C so that many of the contacted cells then underwent hypersensitivity during the next 30 hours. A crude nucleic acid extract from the hypersensitive leaves was made at this time and injected into leaves of new plants of the same wheat line inoculated with a virulent race of rust. The injection caused substantial necrosis of cells which normally remain alive in the presence of a virulent race. In a number of tests, the crude fraction was found to be more effective in causing unusual necrosis in wheat bearing the Sr_6 gene than the sr_6 allele. The specificity of the extract towards lines with the Sr_6 gene was increased by partial purification of the contained RNA, and enzymic destruction of the contained DNA and protein. Specificity was destroyed by RNase treatment, suggesting that it was based on RNA in the extract. However, a marked background activity of the extract remained capable of causing some necrosis non-specifically in either wheat line inoculated with a virulent race. The results of the initial experiments suggested the detection of a nucleic acid as the product of an incompatible relationship involving the avirulence gene P_6 and specifically able to recognize and interact with the resistance gene Sr_6.

Further experiments by Howes, Samborski & Rohringer (1974) implicated the avirulence gene in the parasite as the source of the RNA, and the complementary resistance gene in the host used for the bioassay, as the target. However, it must be noted that the considerable variation from experiment to experiment in the amount of necrosis in the bioassay gives ground for concern in evaluation of the evidence. The criterion for 'activity' was the induction of significantly more necrosis in a wheat line bearing a gene for resistance than in one bearing the recessive allele. Most disturbing was the occasional induction by 'inactive' extracts of large amounts of necrosis in both lines or sometimes in the line bearing the recessive allele. If this problem is disregarded, the avirulence gene in the parasite is seen as the source of the activity, because active extracts were not only obtained from low infection types resulting from the P_6/Sr_6 interaction at 20 °C but also from high infection types resulting from the same interaction at 26 °C and from the P_6/sr_6 interaction, but not from p_6-based interactions. Similarly, an active extract was obtained from the P_5/sr_5 interaction. Activities of RNA from the P_6 and P_5 genes were revealed only in bioassays involving the Sr_6 and Sr_5 genes, respectively, and these were not affected by the virulence or avirulence genes in the races used to inoculate the bioassay plants. It was further argued that haustorial penetration of host cells was necessary for access of the RNA to sites of host gene products. In demonstrating that haustoria of stem rust rather than leaf rust were essential to permit access of the RNA, the authors provided more of the disturbing data referred to above, because they found that leaf rust infection predisposed leaves to exhibit very high levels of necrosis after RNA injection, and most markedly in the line bearing the recessive allele. Thus, this complex and imaginative research effort must be regarded as intriguing but not yet convincing because of the inconsistent results in the bioassay used.

ELICITATION OR SUPPRESSION OF PHYTOALEXIN FORMATION

An obvious means of regulating specificity would be through selective stimulation or suppression of a key defence mechanism in a particular plant. Thus specific control of phytoalexin formation, callose deposition or cork cambium division in a resistant host by a compound secreted by an avirulent parasite should be sought. As discussed in Chapters 5 and 6, many substances are known to be capable of stimulating the production of phytoalexins in legumes and there is some evidence for suppression of phytoalexin formation in potato by components of virulent races of *Phytophthora infestans*. Detection and isolation of molecules of parasitic origin with these types of specific action in the host is largely a problem for future research.

Keen (1975b) found that culture filtrates from avirulent races of *Phytophthora megasperma* var. *sojae* stimulated the accumulation of two to five times more phytoalexin in resistant than susceptible soybean hypocotyls, although there were considerable variations between replicate treatments. In the same experiment, filtrates from a race virulent on both types of hypocotyls caused intermediate amounts of phytoalexin to form. The existence of specific stimulants was suggested by this result, but it is quite clear that much non-specific stimulation was also detected. In fact, at least three fractions with non-specific activity were obtained when culture filtrates of avirulent and virulent races were concentrated and separated by gel filtration. However, filtrates of the avirulent race yielded a fourth fraction which caused several-fold more phytoalexin formation in resistant than susceptible hypocotyls. The possibility that a compound exists with specific ability to cause phytoalexin formation only in resistant soybean awaits further analysis of this particular fraction. Any hypothesis seeking to implicate this specific stimulant in the natural process of infection and host response will have to accommodate the fact that non-specific stimulants exist, and are present in culture filtrates.

CONCLUSION

As stated at the outset of this chapter, no synthesis of view is possible about the existence or nature of recognition factors. Numerous arguments can be made for the existence of macro-molecules involved in mediating specificity during the host–parasite interchange. There are now at least five lines of approach, revealed by the experiments discussed here, which may help in development of hypotheses concerning the ways in which macro-molecules regulate specificity. The significance of toxins active on resistant cells in causing hypersensitivity is easiest to conceive, but evidence in favour of their existence is minimal or lacking. The functions of common antigens as compatibility factors is more difficult to appreciate, yet the evidence for their existence is greater. The discovery of specific cross-protection factors and the rust RNA which affects reactions of cells to parasites may indicate that host and parasite inter-communicate through nucleic acid interchanges, but the evidence is as yet slender. The idea that a parasite may gain specific metabolic control over reactions which normally lead to expression of resistance involving phytoalexin synthesis is attractive, but this is supported by only one set of evidence at present.

References

Albershiem, P., Jones, T. M. & English, P. D. (1969). Biochemistry of the cell wall in relation to infective processes. *A. Rev. Phytopath.* **7**, 171–194.

Allen, E. H. & Thomas, C. A. (1971*a*). Trans-trans-3,11-tridecadiene-5,7,9-triyne-1,2-diol, an antifungal polyacetylene from diseased safflower (*Carthamus tinctorius*). *Phytochemistry* **10**, 1579–1582.

Allen, E. H. & Thomas, C. A. (1971*b*). A second antifungal polyacetylene compound from *Phytophthora*-infected safflower. *Phytopathology* **61**, 1107–1109.

Allen, P. J. (1955). The role of a self-inhibitor in the germination of rust uredospores. *Phytopathology* **45**, 259–266.

Allen, R. F. (1923). A cytological study of infection of Baart and Kanred wheats by *Puccinia graminis tritici*. *J. agric. Res.* **23**, 131–151.

Angell, H. R., Walker, J. C. & Link, K. P. (1930). The relation of protocatechuic acid to disease resistance in the onion. *Phytopathology* **20**, 431–438.

Antonelli, E. & Daly, J. M. (1966). Decarboxylation of indoleacetic acid by near-isogenic lines of wheat resistant and susceptible to *Puccinia graminis* f. sp. *tritici. Phytopathology* **56**, 610–618.

Averre, C. W. & Kelman, A. (1964). Severity of bacterial wilt as influenced by ratio of virulent to avirulent cells of *Pseudomonas solanacearum* in inoculum. *Phytopathology* **54**, 779–783.

Bailey, J. A. (1969). Effects of antimetabolites on production of the phytoalexin pisatin. *Phytochemistry* **8**, 1393–1395.

Bailey, J. A. (1973). Production of antifungal compounds in cowpea (*Vigna sinensis*) and pea (*Pisum sativum*) after virus infection. *J. gen. Microbiol.* **75**, 119–123.

Bailey, J. A. (1974). The relationship between symptom expression and phytoalexin concentration in hypocotyls of *Phaseolus vulgaris* infected with *Colletotrichum lindemuthianum. Physiol. Pl. Path.* **4**, 477–488.

Bailey, J. A. & Burden, R. S. (1973). Biochemical changes and phytoalexin accumulation in *Phaseolus vulgaris* following cellular browning caused by tobacco necrosis virus. *Physiol. Pl. Path.* **3**, 171–177.

Bailey, J. A., Burden, R. S. & Vincent, G. G. (1975). Capsidiol: an antifungal compound produced in *Nicotiana tabacum* and *Nicotiana clevelandii* following infection with tobacco necrosis virus. *Phytochemistry* **14**, 597.

Bailey, J. A. & Deverall, B. J. (1971). Formation and activity of phaseollin in the interaction between bean hypocotyls (*Phaseolus vulgaris*) and physiological races of *Colletotrichum lindemuthianum*. *Physiol. Pl. Path.* **1**, 435–449.

Bailey, J. A. & Ingham, J. L. (1971). Phaseollin accumulation in bean (*Phaseolus vulgaris*) in response to infection by tobacco necrosis virus and the rust *Uromyces appendiculatus*. *Physiol. Pl. Path.* **1**, 451–456.

Beckman, C. H. (1971). The plasticizing of plant cell walls and tylose formation – a model. *Physiol. Pl. Path.* **1**, 1–10.

Berard, D. F., Kuć, J. & Williams, E. B. (1972). A cultivar specific protection factor from incompatible interactions of green bean with *Colletotrichum lindemuthianum*. *Physiol. Pl. Path.* **2**, 123–127.

Berard, D. F., Kuć, J. & Williams, E. B. (1973). Relationship of genes for resistance to protection by diffusates from incompatible interactions of *Phaseolus vulgaris* with *Colletotrichum lindemuthianum*. *Physiol. Pl. Path.* **3**, 51–56.

Bernard, N. (1909). L'évolution dans la symbiose. Les orchidées et leurs champignons commenseaux. *Annls Sci. nat. (Bot.)* **9**, 1–196.

Bernard, N. (1911). Sur la fonction fungicide des bulbes d'Ophrydées. *Annls Sci. nat. (Bot.)* **14**, 221–234.

Blakeman, J. P. & Fraser, A. K. (1971). Inhibition of *Botrytis cinerea* spores by bacteria on the surface of chrysanthemum leaves. *Physiol. Pl. Path.* **1**, 45–54.

Bonde, M. R., Millar, R. L. & Ingham, J. L. (1973). Induction and identification of sativan and vestitol as two phytoalexins from *Lotus corniculatus*. *Phytochemistry* **12**, 2957–2959.

Bracker, C. E. & Littlefield, L. J. (1973). Structural concepts of host–pathogen interfaces. In *Fungal pathogenicity and the plant's response*, pp. 159–313 (ed. R. J. W. Byrde & C. V. Cutting). Academic Press, London & New York.

Brown, A. E. & Swinburne, T. R. (1971). Benzoic acid: an antifungal compound formed in Bramley's Seedling apple fruits following infection by *Nectria galligena* Bres. *Physiol. Pl. Path.* **1**, 469–475.

Brown, A. E. & Swinburne, T. R. (1973). Degradation of benzoic acid by *Nectria galligena* Bres. *in vitro* and *in vivo*. *Physiol. Pl. Path.* **3**, 453–459.

Brown, J. F. & Shipton, W. A. (1964). Relationship of penetration to infection type when seedling wheat leaves are inoculated with *Puccinia graminis tritici*. *Phytopathology* **54**, 89–91.

Brown, J. F., Shipton, W. A. & White, N. H. (1966). The relationship between hypersensitive tissue and resistance in wheat seedlings infected with *Puccinia graminis tritici*. *Ann. appl. Biol.* **58**, 279–290.

Brown, W. (1922). Studies on the physiology of parasitism. 8. On the exosmosis of nutrient substances from the host tissue into the infection drop. *Ann. Bot., Lond.* **36**, 101–119.

Bu'lock, J. D. (1964). Polyacetylenes and related compounds in nature. In *Progress in organic chemistry*, **6**, pp. 86–134 (ed. J. Cook & (W. Carruthers). Butterworths, London.

Burden, R. S. & Bailey, J. A. (1975). Structure of the phytoalexin from soybean. *Phytochemistry* **14**, 1389–1390.

Burden, R. S., Bailey, J. A. & Dawson, G. W. (1972). Structures of three new isoflavonoids from *Phaseolus vulgaris* infected with tobacco necrosis virus. *Tetrahedron Lett. 1972*, 4175–4178.

Burden, R. S., Bailey, J. A. & Vincent, G. G. (1974). Metabolism of phaseollin by *Colletotrichum lindemuthianum. Phytochemistry* **13**, 1789–1791.

Burden, R. S., Bailey, J. A. & Vincent, G. G. (1975). Glutinosone, a new antifungal sesquiterpene from *Nicotiana glutinosa* infected with tobacco mosaic virus. *Phytochemistry* **14**, 221–223.

Buxton, E. W. (1957). Some effects of pea root exudates on physiological races of *Fusarium oxysporum* f. *pisi* (Linf.) Snyder & Hansen. *Trans. Br. mycol. Soc.* **40**, 145–154.

Buxton, E. W. (1962). Root exudates from banana and their relationship to strains of the *Fusarium* causing Panama wilt. *Ann. appl. Biol.* **50**, 269–282.

Carroll, R. B., Lukejic, F. L. & Levine, R. G. (1972). Absence of a common antigen relationship between *Corynebacterium insidiosum* and *Medicago sativa* as a factor in disease development. *Phytopathology* **62**, 1351–1360.

Chester, K. S. (1933). The problem of acquired physiological immunity in plants. *Q. Rev. Biol.* **8**, 119–154 & 275–324.

Chou, M. C. & Preece, T. F. (1968). The effect of pollen grains on infections caused by *Botrytis cinerea* Fr. *Ann. appl. Biol.* **62**, 11–22.

Condon, P. & Kuć, J. (1960). Isolation of a fungitoxic compound from carrot tissue inoculated with *Ceratocystis fimbriata. Phytopathology* **50**, 267–270.

Condon, P. & Kuć, J. (1962). Confirmation of the identity of a fungitoxic compound produced by carrot root tissue. *Phytopathology* **52**, 182–183.

Coxon, D. T., Curtis, R. F., Price, K. R. & Howard, B. (1974). Phytuberin: a novel antifungal terpenoid from potato. *Tetrahedron Lett. 1974*, 2363–2366.

Crosse, J. E. & Garrett, C. M. E. (1970). Pathogenicity of *Pseudomonas morsprunorum* in relation to host specificity. *J. gen. Microbiol.* **62**, 315–327.

Cruickshank, I. A. M. (1962). Studies on phytoalexins. IV. The antimicrobial spectrum of pisatin. *Aust. J. biol. Sci.* **15**, 147–159.

Cruickshank, I. A. M. (1963). Phytoalexins. *A. Rev. Phytopath.* **1**, 351–374.

Cruickshank, I. A. M. & Mandryk, M. (1960). The effect of stem infestation of tobacco with *Peronospora tabacina* on foliage infection to blue mould. *J. Aust. Inst. agric. Sci.* **26**, 369–372.

Cruickshank, I. A. M. & Perrin, D. R. (1960). Isolation of a phytoalexin from *Pisum sativum* L. *Nature, Lond.* **187**, 799–800.

Cruickshank, I. A. M. & Perrin, D. R. (1963a). Phytoalexins of the Leguminosae. Phaseollin from *Phaseolus vulgaris* L. *Life Sci.* **2**, 680–682.

Cruickshank, I. A. M. & Perrin, D. R. (1963b). Studies on phytoalexins. VI. Pisatin: The effect of some factors on its formation in *Pisum sativum* L., and the significance of pisatin in disease resistance. *Aust. J. biol. Sci.* **16**, 111–128.

Cruickshank, I. A. M. & Perrin, D. R. (1965). Studies on phytoalexins. IX. Pisatin formation by cultivars of *Pisum sativum* L. and several other *Pisum* species. *Aust. J. biol. Sci.* **18**, 829–835.

Cruickshank, I. A. M. & Perrin, D. R. (1968). The isolation and partial characterization of monilicolin A, a polypeptide with phaseollin-inducing activity from *Monilinia fructicola*. *Life Sci.* **7**, 449–458.

Cruickshank, I. A. M. & Perrin, D. R. (1971). Studies on phytoalexins. XI. The induction, antimicrobial spectrum and chemical assay of phaseollin. *Phytopath. Z.* **70**, 209–229.

Cruickshank, I. A. M., Veeraraghavan, J. & Perrin, D. R. (1974). Some physical factors affecting the formation and/or net accumulation of medicarpin in infection droplets on white clover leaflets. *Aust. J. Pl. Physiol.* **1**, 149–156.

Daly, J. M. (1972). The use of near-isogenic lines in biochemical studies of the resistance of wheat to stem rust. *Phytopathology* **62**, 392–400.

Davis, D. (1967). Cross-protection in *Fusarium* wilt diseases. *Phytopathology* **57**, 311–314.

Day, P. R. (1974). *Genetics of host-parasite interactions.* W. H. Freeman & Company, San Francisco. 238 pp.

Day, P. R. (1976). Gene function in host–parasite interactions. In *Specificity of plant diseases* (ed. R. K. S. Wood & A. Graniti). Plenum Press, London. (In press.)

Devay, J. E., Charudattan, R. & Wimalajeewa, D. L. S. (1972). Common antigenic determinants as a possible regulation of host–pathogen compatibility. *Am. Nat.* **106**, 185–194.

Deverall, B. J. & Rogers, P. M. (1972). The effect of pH and composition of test solutions on the inhibitory activity of wyerone acid towards germination of fungal spores. *Ann. appl. Biol.* **72**, 301–305.

Deverall, B. J. & Vessey, J. C. (1969). Role of a phytoalexin in controlling lesion development in leaves of *Vicia faba* after infection by *Botrytis* spp. *Ann. appl. Biol.* **63**, 449–458.

Dickinson, S. (1964). The nature of thigmotropic stimuli affecting uredospore germ tubes of *Puccinia* spp. *Trans. Br. mycol. Soc.* **47**, 300.

Dijkman, A. van & Kaars Sijpesteijn, A. (1973). Leakage of pre-absorbed ^{32}P from tomato leaf disks infiltrated with high molecular weight products of incompatible races of *Cladosporium fulvum*. *Physiol. Pl. Path.* **3**, 57–67.

Dineen, J. K. (1963). Antigenic relationships between host and parasite. *Nature, Lond.* **97**, 471–472.

Doubly, J. A., Flor, H. H. & Clagget, C. O. (1960). Relation of antigens of *Melampsora lini* and *Linum usitatissimum* to resistance and susceptibility. *Science* **131**, 229.

Ehrlich, H. G. & Ehrlich, M. A. (1963). Electron microscopy of the host–parasite relationships in stem rust of wheat. *Am. J. Bot.* **50**, 123–130.

Ellingboe, A. H. (1968). Inoculum production and infection by foliage pathogens. *A. Rev. Phytopath.* **6**, 317–330.

Ellingboe, A. H. (1972). Genetics and physiology of primary infection by *Erysiphe graminis. Phytopathology* **62**, 401–413.

Ellingboe, A. H. (1976). Genetics of host–parasite interactions. In *Physiological plant pathology* (ed. P. H. Williams & R. Heitefuss). Springer–Verlag, Berlin. (In press.)

Elliston, J., Kuć, J. & Williams, E. B. (1971). Induced resistance to bean anthracnose at a distance from the site of the inducing interaction. *Phytopathology* **61**, 1110–1112.

Ercolani, G. L. & Crosse, J. E. (1966). The growth of *Pseudomonas phaseolicola* and related plant pathogens *in vivo. J. gen. Microbiol.* **45**, 429–439.

Fawcett, C. H., Firn, R. D. & Spencer, D. M. (1971). Wyerone increase in leaves of broad bean (*Vicia faba* L.) after infection by *Botrytis fabae. Physiol. Pl. Path.* **1**, 163–166.

Flentje, N. T. (1957). Studies on *Pellicularia filamentosa* (Pat.) Rogers. III. Host penetration and resistance and strain specialization. *Trans. Br. mycol. Soc.* **40**, 322–336.

Flentje, N. T., Dodman, R. L. & Kerr, A. (1963). The mechanism of host penetration by *Thanatephorus cucumeris. Aust. J. biol. Sci.* **16**, 784–799.

Flor, H. H. (1956). The complementary genic systems in flax and flax rust. *Adv. Genet.* **8**, 29–54.

Fokkema, N. J. (1968). The influence of pollen on the development of *Cladosporium herbarum* in the phyllosphere of rye. *Neth. J. Pl. Path.* **74**, 159–165.

Fokkema, N. J. (1971). The effect of pollen in the phyllosphere of rye on colonization by saprophytic fungi and on infection by *Helminthosporium sativum* and other leaf pathogens. *Neth. J. Pl. Path.* **77** (suppl. 1), 1–60.

Fries, N. (1966). Chemical factors in the germination of spores of Basidiomycetes. In *The fungus spore,* pp. 189–199 (ed. M. F. Madelin). Butterworths, London.

Garibaldi, A. & Bateman, D. F. (1970). Pectic enzymes produced by *Erwinia chrysanthemi* and their effect on plant tissue. *Physiol. Pl. Path.* **1**, 25–40.

Garrett, C. M. E. & Crosse, J. E. (1975). Interactions between *Pseudomonas morsprunorum* and other pseudomonads in leaf-scar infections of cherry. *Physiol. Pl. Path.* **5**, 89–94.

Gäumann, E. (1964). Weitere Untersuchungen über die chemische Infektabwehr der Orchideen. *Phytopath. Z.* **49**, 211–232.

Gäumann, E. & Jaag, O. (1945). Über induzierte Abwehrreaktionen bei Pflanzen. *Experientia* **1**, 21–22.

Gäumann, E. & Kern, H. (1959a). Über die isolierung und den chemischen Nachweis des Orchinols. *Phytopath. Z.* **35**, 347–356.

Gäumann, E. & Kern, H. (1959b). Über chemische Abwehrreaktionen bei Orchideen. *Phytopath. Z.* **36**, 1–26.

Geigert, J., Stermitz, F. R., Johnson, D., Maag, D. D. & Johnson, D. K.

(1973). Two phytoalexins from sugar beet (*Beta vulgaris*) leaves. *Tetrahedron* **29**, 2703–2706.

Gordon, M., Stoessl, A. & Stothers, J. B. (1973). Post-infectional inhibitors from plants. IV. The structure of capsidiol, an antifungal sesquiterpene from sweet peppers. *Can. J. Chem.* **51**, 748–752.

Grisebach, H. (1965). Biosynthesis of flavonoids. In *Chemistry and biochemistry of plant pigments*, pp. 279–308 (ed. T. W. Goodwin). Academic Press, London.

Grisebach, H. & Hahlbrock, K. (1974). Enzymology and regulation of flavonoid and lignin biosynthesis in plants and plant cell suspension cultures. *Rec. Adv. Phytochem.* **8**, 21–52.

Hadwiger, L. A. (1967). Changes in phenylalanine metabolism associated with pisatin production. *Phytopathology* **57**, 1258–1259.

Hadwiger, L. A., Jafri, A., von Broembsen, S. & Eddy, R. Jr. (1974). Mode of pisatin induction. Increased template activity and dye-binding capacity of chromatin isolated from polypeptide treated pea pods. *Pl. Physiol.* **53**, 52–63.

Hadwiger, L. A. & Schwochau, M. E. (1970). Induction of phenylalanine ammonia lyase and pisatin in pea pods by poly-lysine, spermidine or histone fractions. *Biochem. biophys. Res. Commun.* **38**, 683–691.

Hardegger, E., Biland, H. R. & Corrodi, H. (1963). Synthese von 2,4-Dimethoxy-6-hydroxy-phenanthren und Konstitution des Orchinols. *Helv. chim. Acta* **46**, 1354–1360.

Heath, M. (1974). Light and electron microscope studies, of the interactions of host and non-host plants with cowpea rust – *Uromyces phaseoli* var. *vignae*. *Physiol. Pl. Path.* **4**, 403–414.

Heath, M. C. & Heath, I. B. (1971). Ultrastructure of an immune and a susceptible reaction of cowpea leaves to rust infection. *Physiol. Pl. Path.* **1**, 277–287.

Heath, M. C. & Higgins, V. J. (1973). *In vitro* and *in vivo* conversion of phaseollin and pisatin by an alfalfa pathogen *Stemphylium botryosum*. *Physiol. Plant Path.* **3**, 107–120.

Hecht, E. I. & Bateman, D. F. (1964). Nonspecific acquired resistance to pathogens resulting from localized infections by *Thielaviopsis basicola* or viruses in tobacco leaves. *Phytopathology* **54**, 523–530.

Hess, S. L., Hadwiger, L. A. & Schwochau, M. E. (1971). Studies on biosynthesis of phaseollin in excised pods of *Phaseolus vulgaris*. *Phytopathology* **61**, 79–82.

Heuvel, J. van den & Van Etten, H. D. (1973). Detoxification of phaseollin by *Fusarium solani* f. sp. *phaseoli*. *Physiol. Pl. Path.* **3**, 327–339.

Heuvel, J. van den, Van Etten, H. D., Serum, J. W., Coffen, D. L. & Williams, T. H. (1974). Identification of 1α-hydroxyphaseollone, a phaseollin metabolite produced by *Fusarium solani*. *Phytochemistry* **13**, 1129–1131.

Hickman, C. J. & Ho, H. H. (1966). Behaviour of zoospores in plant-pathogenic Phycomycetes. *A. Rev. Phytopath.* **4**, 195–220.

Hietala, P. K. & Wahlroos, O. (1956). The synthesis of 6-methoxy-2(3)-benzoxazolinone. *Acta chem. scand.* **10**, 1196–1197.

Higgins, V. J. (1972). Role of the phytoalexin medicarpin in three leaf spot diseases of alfalfa. *Physiol. Pl. Path.* **2**, 289–300.

Higgins, V. J. (1975). Induced conversion of the phytoalexin maackiain to dihydromaackiain by the alfalfa pathogen *Stemphylium botryosum.* *Physiol. Pl. Path.* **6**, 5–18.

Higgins, V. J. & Millar, R. L. (1968). Phytoalexin production by alfalfa in response to infection by *Colletotrichum phomoides, Helminthosporium turcicum, Stemphylium loti,* and *S. botryosum. Phytopathology* **58**, 1377–1383.

Higgins, V. J. & Millar, R. L. (1969a). Comparative abilities of *Stemphylium botryosum* and *Helminthosporium turcicum* to induce and degrade a phytoalexin from alfalfa. *Phytopathology* **59**, 1493–1499.

Higgins, V. J. & Millar, R. L. (1969b). Degradation of alfalfa phytoalexin by *Stemphylium botryosum. Phytopathology* **59**, 1500–1506.

Higgins, V. J. & Smith, D. G. (1972). Separation and identification of two pterocarpanoid phytoalexins produced by red clover leaves. *Phytopathology* **62**, 235–238.

Higgins, V. J., Stoessl, A. & Heath, M. C. (1974). Conversion of phaseollin to phaseollinisoflavan by *Stemphylium botryosum. Phytopathology* **64**, 105–107.

Hiura, M. (1943). Studies in storage and rot of sweet potato. *Sci. Rep. Gifu agric. Coll. Japan* **50**, 1–5.

Howes, N. K., Samborski, D. J. & Rohringer, R. (1974). Production and bioassay of gene-specific RNA determining resistance of wheat to stem rust. *Can. J. Bot.* **52**, 2489–2497.

Hsu, S. C. & Ellingboe, A. H. (1972). Elongation of secondary hyphae and transfer of ³⁵S from barley to *Erysiphe graminis* f.sp. *hordei* during primary infection. *Phytopathology* **62**, 876–882.

Hutchinson, J. (1964). *The genera of flowering plants,* **1** *Dicotyledons.* Clarendon, Oxford.

Ingham, J. L. (1972). Phytoalexins and other natural products as factors in plant disease resistance. *Bot. Rev.* **38**, 343–424.

Ingham, J. L. (1973). Disease resistance in higher plants. The concept of pre-infectional and post-infectional resistance. *Phytopath. Z.* **78**, 314–335.

Ingham, J. L. & Millar, R. L. (1973). Sativin: an induced isoflavan from the leaves of *Medicago sativa* L. *Nature, Lond.* **242**, 125–126.

Johnson, C., Brannon, D. R. & Kuć, J. (1973). Xanthotoxin. Phytoalexin of *Pastinaca sativa* root. *Phytochemistry* **12**, 2961–2962.

Johnson, R. (1976a). Genetics of fungal–plant interactions. In *Biochemical aspects of plant parasite relationships* (ed. J. Friend). Academic Press, London. (In press.)

Johnson, R. (1976b). Genetics of host–parasite interactions. In *Specificity in plant diseases* (ed. R. K. S. Wood & A. Graniti). Plenum Press, London. (In press.)

Johnston, C. O. & Huffman, M. D. (1958). Evidence of local antagonism between two cereal rust fungi. *Phytopathology* **48**, 69–70.

Jones, D. R., Graham, W. G. & Ward, E. W. B. (1974). Ultrastructural changes in pepper cells in a compatible interaction with *Phytophthora capsici*. *Phytopathology* **64**, 1084–1090.

Jones, D. R., Graham, W. G. & Ward, E. W. B. (1975a). Ultrastructural changes in pepper cells in an incompatible interaction with *Phytophthora infestans*. *Phytopathology* **65**, 1274–1285.

Jones, D. R., Graham, W. G. & Ward, E. W. B. (1975b). Ultrastructural changes in pepper cells in interactions with *Phytophthora capsici* (isolate 18) and *Monilinia fructicola*. *Phytopathology* **65**, 1409–1416.

Katsui, N., Murai, A., Takasugi, M., Imaizumi, K., Masumane, T. & Tomiyama, K. (1968). The structure of rishitin, a new antifungal compound from diseased potato tubers. *Chem. Commun. 1968*, 43–44.

Keen, N. T. (1971). Hydroxyphaseollin production by soybeans resistant and susceptible to *Phytophthora megasperma* var. *sojae*. *Physiol. Pl. Path.* **1**, 265–275.

Keen, N. T. (1972). Accumulation of wyerone in broad bean and demethylhomopterocarpin in jack bean following inoculation with *Phytophthora megasperma* var. *sojae*. *Phytopathology* **62**, 1365.

Keen, N. T. (1975a). Isolation of phytoalexins from germinating seeds of *Cicer arietinum*, *Vigna sinensis*, *Arachis hypogaea* and other plants. *Phytopathology* **65**, 91–92.

Keen, N. T. (1975b). Specific elicitors of plant phytoalexin production: determinants of race specificity in pathogens? *Science* **187**, 74–75.

Keen, N. T. & Kennedy, B. W. (1974). Hydroxyphaseollin and related isoflavanoids in the hypersensitive resistance reaction of soybeans to *Pseudomonas glycinea*. *Physiol. Pl. Path.* **4**, 173–185.

Kerr, A. (1972). Biological control of crown gall: seed inoculation. *J. appl. Bact.* **35**, 493–497.

Kerr, A. & Flentje, N. T. (1957). Host infection in *Pellicularia filamentosa* controlled by chemical stimuli. *Nature, Lond.* **179**, 204–205.

Kerr, A. & Htay, K. (1974). Biological control of crown gall through bacteriocin production. *Physiol. Pl. Path.* **4**, 37–44.

Khew, K. L. & Zentmyer, G. A. (1973). Chemotactic response of zoospores of five species of *Phytophthora*. *Phytopathology* **63**, 1511–1517.

Khew, K. L. & Zentmyer, G. A. (1974). Electrotactic response of zoospores of seven species of *Phytophthora*. *Phytopathology* **64**, 500–507.

King, L., Hampton, R. E. & Diachun, S. (1964). Resistance to *Erysiphe polygoni* of red clover infected with bean yellow mosaic virus. *Science* **146**, 1054–1055.

Kiraly, Z., Barna, B. & Ersek, T. (1972). Hypersensitivity as a consequence, not the cause, of plant resistance to infection. *Nature, Lond.* **239**, 456–458.

Kitazawa, K. & Tomiyama, K. (1969). Microscopic observations of infection of potato cells by compatible and incompatible races of *Phytophthora infestans*. *Phytopath. Z.* **66**, 317–324.

Klement, Z. & Goodman, R. N. (1967). The hypersensitive reaction to infection by bacterial plant pathogens. *A. Rev. Phytopath.* **5**, 17–44.

Kochman, J. K. & Brown, J. F. (1975). Studies on the mechanism of cross-protection in cereal rusts. *Physiol. Pl. Path.* **6**, 19–27.

Kommedahl, T. (1966). Relation of exudates of pea roots to germination of spores in races of *Fusarium oxysporum* f. *pisi*. *Phytopathology* **56**, 721–722.

Kubota, T. & Matsuura, T. (1953). Chemical studies on the black rot disease of sweet potato. *J. chem. Soc. Japan (Pure Chem. Sect.)* **74**, 248–251.

Kuć, J. (1976). Terpenoid phytoalexins. In *Biochemical aspects of plant parasite relationships* (ed. J. Friend). Academic Press, London. (In press.)

Kuć, J., Shockley, G. & Kearney, K. (1975). Protection of cucumber against *Colletotrichum lagenarium* by *Colletotrichum lagenarium*. *Physiol. Pl. Path.* **7**, 195–199.

Kuo, M. S., Yoder, O. C. & Scheffer, R. P. (1970). Comparative specificity of the toxins of *Helminthosporium carbonum* and *Helminthosporium victoriae*. *Phytopathology* **60**, 365–368.

Lesemann, D. E. & Fuchs, W. H. (1970). Die Ultrastruktur des Penetrations-vorganges von *Olpidium brassicae* an Kohlrabi-Wurzeln. *Archs Mikrobiol.* **71**, 20–30.

Letcher, R. M., Widdowson, D. A., Deverall, B. J. & Mansfield, J. W. (1970). Identification and activity of wyerone acid as a phytoalexin in broad bean (*Vicia faba*) after infection by *Botrytis*. *Phytochemistry* **9**, 249–252.

Link, K. P., Dickson, A. D. & Walker, J. C. (1929). Further observations on the occurrence of protocatechuic acid in pigmented onion scales and its relation to disease resistance in the onion. *J. biol. Chem.* **84**, 719–725.

Link, K. P. & Walker, J. C. (1933). The isolation of catechol from pigmented onion scales and its significance in relation to disease resistance in onions. *J. biol. Chem.* **100**, 379–383.

Lippincott, B. B. & Lippincott, J. A. (1969). Bacterial attachment to a specific wound site as an essential stage in tumor initiation by *Agrobacterium tumefaciens*. *J. Bact.* **97**, 620–628.

Littlefield, L. J. (1969). Flax rust resistance induced by prior inoculation with an avirulent race of *Melampsora lini*. *Phytopathology* **59**, 1323–1328.

Loebenstein, G. & Lovrekovich, L. (1966). Interference with tobacco mosaic virus local lesion formation in tobacco by injecting heat-killed cells of *Pseudomonas syringae*. *Virology* **30**, 587–591.

Loegering, W. Q. (1966). The relationship between host and pathogen in stem rust of wheat. *Proceedings of the 2nd International Wheat Genetics Symposium* (Lund, 1963). *Hereditas*, Suppl. 2, 167–177.

Lozano, J. C. & Sequeira, L. (1970). Prevention of the hypersensitive reaction in tobacco leaves by heat-killed bacterial cells. *Phytopathology* **60**, 875–879.

Lyon, F. M. & Wood, R. K. S. (1975). Production of phaseollin, coumestrol and related compounds in bean leaves inoculated with *Pseudomonas* spp. *Physiol. Pl. Path.* **6**, 117–124.

Lyon, G. D. (1972). Occurrence of rishitin and phytuberin in potato tubers inoculated with *Erwinia carotovora* var. *atroseptica*. *Physiol. Pl. Path.* **2**, 411–416.

Lyon, G. D. & Bayliss, C. E. (1975). The effect of rishitin on *Erwinia carotovora* var. *atroseptica* and other bacteria. *Physiol. Pl. Path.* **6**, 177–186.

Lyon, G. D., Lund, B. M., Bayliss, C. E. & Wyatt, G. H. (1975). Resistance of potato tubers to *Erwinia carotovora* and formation of rishitin and phytuberin in infected tissue. *Physiol. Pl. Path.* **6**, 43–50.

Macko, V., Staples, R. C., Allen, P. J. & Renwick, J. A. A. (1971). Identification of the germination self-inhibitor from wheat stem rust uredospores. *Science* **173**, 835–836.

Macko, V., Staples, R. C., Renwick, J. A. A. & Pirone, J. (1972). Germination self-inhibitors of rust uredospores. *Physiol. Pl. Path.* **2**, 347–355.

Maclean, D. J., Sargent, J. A., Tommerup, I. C. & Ingram, D. S. (1974). Hypersensitivity as the primary event in resistance to fungal parasites. *Nature, Lond.* **249**, 186–187.

Magrou, J. (1924). L'immunité humorale chez les plantes. *Rev. Path. Vég. Entom. Agr.* **11**, 189–192.

Maheshwari, R., Allen, P. J. & Hildebrand, A. C. (1967). Physical and chemical factors controlling the development of infection structures from urediospore germ tubes of rust fungi. *Phytopathology* **57**, 855–862.

Mansfield, J. W. & Deverall, B. J. (1971). Mode of action of pollen in breaking resistance of *Vicia faba* to *Botrytis cinerea*. *Nature, Lond.* **232**, 339.

Mansfield, J. W. & Deverall, B. J. (1974a). The rates of fungal development and lesion formation in leaves of *Vicia faba* during infection by *Botrytis cinerea* and *Botrytis fabae*. *Ann. appl. Biol.* **76**, 77–89.

Mansfield, J. W. & Deverall, B. J. (1974b). Changes in wyerone acid concentrations in leaves of *Vicia faba* after infection by *Botrytis cinerea* or *B. fabae*. *Ann. appl. Biol.* **77**, 227–235.

Mansfield, J. W., Hargreaves, J. A. & Boyle, F. C. (1974). Phytoalexin production by live cells in broad bean leaves infected with *Botrytis cinerea*. *Nature, Lond.* **252**, 316–317.

Mansfield, J. W. & Widdowson, D. A. (1973). The metabolism of wyerone acid (a phytoalexin from *Vicia faba* L.) by *Botrytis fabae* and *B. cinerea*. *Physiol. Pl. Path.* **3**, 393–404.

Marks, G. C., Berbee, J. G. & Riker, A. J. (1965). Direct penetration of leaves of *Populus tremuloides* by *Colletotrichum gloeosporioides*. *Phytopathology* **55**, 408–412.

Martin, J. T. (1964). Role of cuticle in the defense against plant disease. *A. Rev. Phytopath.* **2**, 81–100.

Martin, J. T. & Juniper, B. E. (1970). *The cuticles of plants*. Edward Arnold, London. 347 pp.

Martin, T. J., Stuckey, R. E., Safir, G. R. & Ellingboe, A. H. (1975). Reduction of transpiration from wheat caused by germinating conidia of *Erysiphe graminis* f. sp. *tritici*. *Physiol. Pl. Path.* **7**, 71–77.

Matthews, R. E. F. (1970). *Plant virology*. Academic Press, New York & London. 778 pp.

Mayama, S., Daly, J. M., Rehfeld, D. W. & Daly, C. R. (1975). Hypersensitive response of near-isogenic wheat carrying the temperature-sensitive Sr_6 allele for resistance to stem rust. *Physiol. Pl. Path.* **7**, 35–47.

McIntyre, J. L., Kuć, J. & Williams, E. B. (1975). Protection of Bartlett pear against fire blight with deoxyribonucleic acid from virulent and avirulent *Erwinia amylovora*. *Physiol. Pl. Path.* **7**, 153–170.

McKeen, W. E. (1974). Mode of penetration of epidermal cell walls of *Vicia faba* by *Botrytis cinerea*. *Phytopathology* **64**, 461–467.

Meehan, F. L. & Murphy, H. C. (1946). A new *Helminthosporium* blight of oats. *Science* **104**, 413–414.

Meehan, F. L. & Murphy, H. C. (1947). Differential phytotoxicity of metabolic by-products of *Helminthosporium victoriae*. *Science* **106**, 270–271.

Mercer, P. C., Wood, R. K. S. & Greenwood, A. D. (1974). Resistance to anthracnose of French bean. *Physiol. Pl. Path.* **4**, 291–306.

Metlitskii, L. V., Ozeretskovskaya, O. L., Vul'fson, N. S. & Chalova, L. I. (1971). Chemical nature of lubimin, a new phytoalexin of potatoes. *Dokl. Akad. Nauk SSSR* **200**, 1470–1472.

Müller, K. O. (1958). Studies on phytoalexins. I. The formation and immunological significance of phytoalexin produced by *Phaseolus vulgaris* in response to infections with *Sclerotinia fructicola* and *Phytophthora infestans*. *Aust. J. biol. Sci.* **11**, 275–300.

Müller, K. O. (1959). Hypersensitivity. In *Plant pathology*, **1**, pp. 469–519 (ed. J. G. Horsfall & A. E. Dimond). Academic Press, New York.

Müller, K. O. & Börger, H. (1941). Experimentelle Untersuchungen über die *Phytophthora* – Resistenz der Kartoffel. *Arb. biol. Anst. (Reichsaust) Berl.* **23**, 189–231.

New, P. B. & Kerr, A. (1972). Biological control of crown gall: field measurements and glasshouse experiments. *J. appl. Bact.* **35**, 279–287.

Nobécourt, P. (1923). Sur la production d'anticorp par les tubercules des Ophrydées. *C.r. Acad. Sci. Paris* **177**, 1055–1057.

Nusbaum, C. J. & Keitt, G. W. (1938). A cytological study of host–parasite relations of *Venturia inaequalis* on apple leaves. *J. agric. Res.* **56**, 595–618.

Ogle, H. & Brown, J. F. (1971). Quantitative studies of the post-penetration phase of infection by *Puccinia graminis tritici*. *Ann. appl. Biol.* **67**, 309–319.

Oguni, I. & Uritani, I. (1974). Dehydroipomeamarone as an intermediate in the biosynthesis of ipomeamarone, a phytoalexin from sweet potato root infected with *Ceratocystis fimbriata*. *Pl. Physiol.* **53**, 649–652.

Olah, A. F. & Sherwood, R. T. (1971). Flavones, isoflavones, and coumestans in alfalfa infected by *Ascochyta imperfecta*. *Phytopathology* **61**, 65–69.

Olah, A. F. & Sherwood, R. T. (1973). Glycosidase activity and flavanoid

99

accumulation in alfalfa infected by *Ascochyta imperfecta. Phytopathology* **63**, 739–742.

Paxton, J., Goodchild, D. J. & Cruickshank, I. A. M. (1974). Phaseollin production by live bean endocarp. *Physiol. Pl. Path.* **4**, 167–171.

Perrin, D. R. (1964). The structure of phaseollin. *Tetrahedron Lett. 1964*, 29–35.

Perrin, D. R. & Bottomley, W. (1962). Studies on phytoalexins. V. The structure of pisatin from *Pisum sativum* L. *J. Am. chem. Soc.* **84**, 1919–1922.

Perrin, D. R. & Cruickshank, I. A. M. (1965). Studies on phytoalexins. VII. Chemical stimulation of pisatin formation in *Pisum sativum* L. *Aust. J. biol. Sci.* **18**, 803–816.

Perrin, D. R., Whittle, C. P. & Batterham, T. J. (1972). The structure of phaseollidin. *Tetrahedron Lett. 1972*, 1673–1676.

Pierre, R. E. & Millar, R. L. (1965). Histology of pathogen–suscept relationship of *Stemphylium botryosum* and alfalfa. *Phytopathology* **55**, 909–914.

Pont, W. (1959). Blue mould (*Peronospora tabacina* Adam) of tobacco in North Queensland: some aspects of chemical control. *Q. J. agric. Sci.* **16**, 299–327.

Preece, T. F. (1963). Micro-exploration and mapping of apple scab infections. *Trans. Br. mycol. Soc.* **46**, 523–529.

Preece, T. F. & Dickinson, C. H. (1971). *Ecology of leaf surface microorganisms.* Academic Press, London & New York. 640 pp.

Pringle, R. B. (1972). Chemistry of host-specific phytotoxins. In *Phytotoxins in plant diseases*, pp. 139–154 (ed. R. K. S. Wood, A. Ballio & A. Graniti). Academic Press, London & New York.

Pringle, R. B. & Braun, A. C. (1957). The isolation of the toxin of *Helminthosporium victoriae. Phytopathology* **47**, 369–371.

Pueppke, S. G. & Van Etten, H. D. (1974). Pisatin accumulation and lesion development in peas infected with *Aphanomyces euteiches, Fusarium solani* f. sp. *pisi,* or *Rhizoctonia solani. Phytopathology* **64**, 1433–1440.

Rahe, J. E. (1973). Occurrence and levels of the phytoalexin phaseollin in relation to delimitation at sites of infection of *Phaseolus vulgaris* by *Colletotrichum lindemuthianum. Can. J. Bot.* **51**, 2423–2430.

Rahe, J. E. & Arnold, R. M. (1975). Injury-related phaseollin accumulation in *Phaseolus vulgaris* and its implications with regard to specificity of host–parasite interaction. *Can. J. Bot.* **53**, 921–928.

Rathmell, W. G. (1973). Phenolic compounds and phenylalanine ammonia lyase activity in relation to phytoalexin biosynthesis in infected hypocotyls of *Phaseolus vulgaris. Physiol. Pl. Path.* **3**, 259–267.

Rathmell, W. G. & Bendall, D. S. (1972). The peroxidase-catalysed oxidation of a chalcone and its possible physiological significance. *Biochem. J.* **127**, 125–132.

Roberts, M. F., Martin, J. T. & Peries, O. S. (1961). Studies on plant cuticles. IV. The leaf cuticle in relation to invasion by fungi. *A. Rep. Long Ashton Res. Sta. 1960*, 102–110.

Rohde, R. A. & Jenkins, W. R. (1958). The chemical basis for resistance of asparagus to the nematode *Trichodorus christiei*. *Phytopathology* **48**, 463.

Rohringer, R., Howes, N. K., Kim, W. K. & Samborski, D. J. (1974). Evidence for a gene-specific RNA determining resistance in wheat to stem rust. *Nature, Lond.* **249**, 585–587.

Ross, A. F. & Israel, H. W. (1970). Use of heat treatments in the study of acquired resistance to tobacco mosaic virus in hypersensitive tobacco. *Phytopathology* **60**, 755–770.

Rovira, A. D. (1965). Plant root exudates and their influence upon soil microorganisms. In *Ecology of soil-borne plant pathogens*, pp. 170–184 (ed. K. F. Baker & W. C. Snyder). Univ. Calif. Press, Berkeley.

Royle, D. J. & Hickman, C. J. (1964a). Analysis of factors governing *in vitro* accumulation of zoospores of *Pythium aphanidermatum* on roots. I. Behaviour of zoospores. *Can. J. Microbiol.* **10**, 151–162.

Royle, D. J. & Hickman, C. J. (1964b). Analysis of factors governing *in vitro* accumulation of zoospores of *Pythium aphanidermatum* on roots. II. Substances causing response. *Can. J. Microbiol.* **10**, 201–219.

Royle, D. J. & Thomas, G. G. (1971). The influence of stomatal opening on the infection of hop leaves by *Pseudoperonospora humuli*. *Physiol. Pl. Path.* **1**, 329–343.

Royle, D. J. & Thomas, G. G. (1973). Factors affecting zoospore responses towards stomata in hop downy mildew (*Pseudoperonospora humuli*) including some comparisons with grapevine downy mildew (*Plasmopara viticola*). *Physiol. Pl. Path.* **3**, 405–417.

Sato, N., Tomiyama, K., Katsui, N. & Masamune, T. (1968). Isolation of rishitin from tomato plants. *Ann. Phytopath. Soc. Japan* **34**, 344–345.

Scheffer, R. P. & Pringle, R. B. (1963). Respiratory effects of the selective toxin of *Helminthosporium victoriae*. *Phytopathology* **53**, 465–468.

Scheffer, R. P. & Yoder, O. C. (1972). Host-specific toxins and selective toxicity. In *Phytotoxins in plant diseases*, pp. 251–269 (ed. R. K. S. Wood, A. Ballio & A. Graniti). Academic Press, London & New York.

Schnathorst, W. C. & Devay, J. E. (1963). Common antigens in *Xanthomonas malvacearum* and *Gossypium hirsutum* and their possible relationship to host specificity and disease resistance. *Phytopathology* **53**, 1142 (abstr.).

Schroth, M. N. & Hildebrand, D. C. (1964). Influence of plant exudates on root-infecting fungi. *A. Rev. Phytopath.* **2**, 101–132.

Sequeira, L., Aist, S. & Ainslie, V. (1972). Prevention of the hypersensitive reaction in tobacco by proteinaceous constituents of *Pseudomonas solanacearum*. *Phytopathology* **62**, 536–541.

Shih, M. & Kuć, J. (1973). Incorporation of ^{14}C from acetate and mevalonate into rishitin and steroid glycoalkaloids by potato tuber slices inoculated with *Phytophthora infestans*. *Phytopathology* **63**, 826–829.

Shih, M., Kuć, J. & Williams, E. B. (1973). Suppression of steroid glycoalkaloid accumulation as related to rishitin accumulation in potato tubers. *Phytopathology* **63**, 821–826.

Shimony, C. & Friend, J. (1975). Ultrastructure of the interaction between *Phytophthora infestans* and leaves of two cultivars of potato (*Solanum tuberosum* L.). Orion and Majestic. *New Phytol.* **74**, 59–65.

de Silva, R. L. & Wood, R. K. S. (1964). Infection of plants by *Corticium solani* and *C. praticola* – effect of plant exudates. *Trans. Br. mycol. Soc.* **47**, 15–24.

Sims, J. J., Keen, N. T. & Honwad, V. K. (1972). Hydroxyphaseollin, an induced antifungal compound from soybeans. *Phytochemistry* **11**, 827–828.

Skipp, R. A. & Deverall, B. J. (1972). Relationships between fungal growth and host changes visible by light microscopy during infection of bean hypocotyls (*Phaseolus vulgaris*) susceptible and resistant to physiologic races of *Colletotrichum lindemuthianum*. *Physiol. Pl. Path.* **2**, 357–374.

Skipp, R. A. & Deverall, B. J. (1973). Studies on cross-protection in the anthracnose disease of bean. *Physiol. Pl. Path.* **3**, 299–314.

Skipp, R. A. & Samborski, D. J. (1974). The effect of the Sr_6 gene for host resistance on histological events during the development of stem rust in near-isogenic wheat lines. *Can. J. Bot.* **52**, 1107–1115.

Slesinski, R. S. & Ellingboe, A. H. (1969). The genetic control of primary infection of wheat by *Erysiphe graminis* f. sp. *tritici*. *Phytopathology* **59**, 1833–1837.

Slesinski, R. S. & Ellingboe, A. H. (1970). Gene-for-gene interactions during primary infection of wheat by *Erysiphe graminis* f. sp. *tritici*. *Phytopathology* **60**, 1068–1070.

Slesinski, R. S. & Ellingboe, A. H. (1971). Transfer of ^{35}S from wheat to the powdery mildew fungus with compatible and incompatible parasite/host genotypes. *Can. J. Bot.* **49**, 303–310.

Smith, A. M. (1973). Ethylene as a cause of soil fungistasis. *Nature, Lond.* **246**, 311–313.

Smith, A. M. & Cook, R. J. (1974). Implications of ethylene production by bacteria for biological balance of soil. *Nature, Lond.* **252**, 703–705.

Smith, D. A., Van Etten, H. D. & Bateman, D. F. (1975). Accumulation of phytoalexins in *Phaseolus vulgaris* hypocotyls following infection by *Rhizoctonia solani*. *Physiol. Pl. Path.* **5**, 51–64.

Smith, D. A., Van Etten, H. D., Serum, J. W., Jones, T. M., Bateman, D. F., Williams, T. H. & Coffen, D. L. (1973). Confirmation of the structure of kievitone, an antifungal isoflavanone isolated from *Rhizoctonia*-infected bean tissues. *Physiol. Pl. Path.* **3**, 293–297.

Smith, D. G., McInnes, A. G., Higgins, V. J. & Millar, R. L. (1971). Nature of the phytoalexin produced by alfalfa in response to fungal infection. *Physiol. Pl. Path.* **1**, 41–44.

Smith, O. F. (1938). Host–parasite relations in red clover plants resistant and susceptible to powdery mildew, *Erysiphe polygoni*. *J. agric. Res.* **57**, 671–682.

Sörensen, N. A. (1961). Some naturally occurring acetylenic compounds. *Chem. Soc. (Lond.) Proc. 1961*, 98–110.

Stakman, E. C. (1915). Relation between *Puccinia graminis* and plants highly resistant to its attack. *J. agric. Res.* **4**, 193–200.

Steiner, G. W. & Byther, R. S. (1971). Partial characterization and use of a host specific toxin from *Helminthosporium sacchari* on sugarcane. *Phytopathology* **61**, 691–695.

Steiner, P. W. & Millar, R. L. (1974). Degradation of medicarpin and sativan by *Stemphylium botryosum*. *Phytopathology* **64**, 586 (abstr.).

Stephens, G. J. & Wood, R. K. S. (1975). Killing of protoplasts by soft-rot bacteria. *Physiol. Pl. Path.* **5**, 165–181.

Stholasuta, P., Bailey, J. A., Severin, V. & Deverall, B. J. (1971). Effect of bacterial inoculation of bean and pea leaves on the accumulation of phaseollin and pisatin. *Physiol. Pl. Path.* **1**, 177–184.

Stoessl, A. (1972). Inermin associated with pisatin in peas inoculated with the fungus *Monilinia fructicola*. *Can. J. Biochem.* **50**, 107–108.

Stoessl, A., Unwin, C. H. & Ward, E. W. B. (1973). Postinfectional fungus inhibitors from plants: fungal oxidation of capsidiol in pepper fruit. *Phytopathology* **63**, 1225–1231.

Strange, R. N., Majer, J. R. & Smith, H. (1974). The isolation and identification of choline and betaine as the two major components in anthers and wheat germ that stimulate *Fusarium graminearum in vitro*. *Physiol. Pl. Path.* **4**, 277–290.

Strange, R. N. & Smith, H. (1971). A fungal growth stimulant in anthers which predisposes wheat to attack by *Fusarium graminearum*. *Physiol. Pl. Path.* **1**, 141–150.

Strobel, G. A. (1973*a*). Biochemical basis of the resistance of sugarcane to eyespot disease. *Proc. natn. Acad. Sci. USA* **70**, 1693–1696.

Strobel, G. A. (1973*b*). Helminthosporoside-binding protein of sugarcane. *J. biol. Chem.* **248**, 1321–1328.

Strobel, G. A. & Hess, W. M. (1974). Evidence for the presence of the toxin-binding protein on the plasma membrane of sugarcane cells. *Proc. natn. Acad. Sci. USA* **71**, 1413–1417.

Stuckey, R. E. & Ellingboe, A. H. (1974). Elongation of secondary hyphae of *Erysiphe graminis* f. sp. *tritici* on wheat with compatible and incompatible parasite/host genotypes. *Phytopathology* **64**, 530–533.

Sussman, A. S. (1966). Dormancy and spore germination. In *The fungi*, **2** *The fungal organism*, pp. 733–764 (ed. G. C. Ainsworth & A. S. Sussman). Academic Press, New York & London.

Sztejnberg, A. & Blakeman, J. P. (1973). Ultraviolet–induced changes in populations of epiphytic bacteria on beetroot leaves and their effect on germination of *Botrytis cinerea* spores. *Physiol. Pl. Path.* **3**, 443–451.

Talboys, P. W. (1972). Resistance to vascular wilt fungi. *Proc. R. Soc. Lond.* B **181**, 319–332.

Tomiyama, K. (1955). Cytological studies on resistance of potato plants to *Phytophthora infestans*. 2. The death of the intracellular hypha in the hypersensitive cell. *Ann. Phytopath. Soc. Japan* **19**, 149–154.

Tomiyama, K. (1967). Further observations on the time requirement for hypersensitive cell death of potatoes infected by *Phytophthora infestans* and its relation to metabolic activity. *Phytopath. Z.* **58**, 367–378.

Tomiyama, K., Sakuma, T., Ishizaka, N., Sato, N., Katsui, N., Takasugi,

M. & Masamune, T. (1968). A new antifungal substance isolated from resistant potato tuber tissue infected by pathogens. *Phytopathology* **58**, 115–116.

Tribe, H. T. (1955). Studies in the physiology of parasitism. XIX. On the killing of plant cells by enzymes from *Botrytis cinerea* and *Bacterium aroideae*. *Ann. Bot.* **19**, 351–371.

Trione, E. J. (1950). The HCN content of flax in relation to flax wilt resistance. *Phytopathology* **50**, 482–486.

Troutman, J. L. & Wills, W. H. (1964). Electrotaxis of *Phytophthora parasitica* zoospores and its possible role in infection of tobacco by the fungus. *Phytopathology* **54**, 225–228.

Turner, J. G. & Novacky, A. (1974). The quantitative relation between plant and bacterial cells involved in the hypersensitive reaction. *Phytopathology* **64**, 885–890.

Van Etten, H. D. & Pueppke, S. G. (1976). Isoflavonoid Phytoalexins. In *Biochemical aspects of plant parasite relationships* (ed. J. Friend). Academic Press, London. (In press.)

Van Etten, H. D. & Smith, D. A. (1975). Accumulation of antifungal isoflavanoids and 1α-hydroxyphaseollone, a phaseollin metabolite, in bean tissue infected with *Fusarium solani* f. sp. *phaseoli*. *Physiol. Pl. Path.* **5**, 225–237.

Varns, J. L., Currier, W. W. & Kuć, J. (1971). Specificity of rishitin and phytuberin accumulation by potato. *Phytopathology* **61**, 968–971.

Varns, J. L. & Kuć, J. (1971). Suppression of rishitin and phytuberin accumulation and hypersensitive response in potato by compatible races of *Phytophthora infestans*. *Phytopathology* **61**, 178–181.

Varns, J. L., Kuć, J. & Williams, E. B. (1971). Terpenoid accumulation as a biochemical response of the potato tuber to *Phytophthora infestans*. *Phytopathology* **61**, 174–177.

Virtanen, A. I. & Hietala, P. K. (1959). On the structure of the precursors of benzoxazolinone in rye plants. *Suom. Kemistilehti* **32**(B), 252.

Wacek, T. J. & Sequeira, L. (1973). The peptidoglycan of *Pseudomonas solanacearum*: chemical composition and biological activity in relation to the hypersensitive reaction in tobacco. *Physiol. Pl. Path.* **3**, 363–369.

Wahlroos, O. & Virtanen, A. I. (1959). The precursors of 6-methoxybenzoxazolinone in maize and wheat plants, their isolation and some of their properties. *Acta chem. scand.* **13**, 1906–1908.

Wallace, H. R. (1973). *Nematode ecology and plant disease*. Edward Arnold, London. 228 pp.

Ward, E. W. B., Unwin, C. H. & Stoessl, A. (1973). Post-infectional inhibitors from plants. VI. Capsidiol production in pepper fruit infected with bacteria. *Phytopathology* **63**, 1537–1538.

Ward, H. M. (1905). Recent researches on the parasitism of fungi. *Ann. Bot.* **19**, 1–54.

Whalley, W. M. & Taylor, G. S. (1973). Influence of pea-root exudates on germination of conidia and chlamydospores of physiologic races of *Fusarium oxysporum* f. *pisi*. *Ann. appl. Biol.*, **73**, 269–276.

Williams, P. H., Aist, J. R. & Bhattacharya, P. K. (1973). Host–parasite relations in cabbage clubroot. In *Fungal pathogenicity and the host's response*, pp. 141–155 (ed. R. J. W. Byrde & C. V. Cutting). Academic Press, London & New York.

Williams, P. H. & McNabola, S. S. (1970). Fine structure of the host–parasite interface of *Plasmodiophora brassicae* in cabbage. *Phytopathology* **60**, 1557–1561.

Wimalajeewa, D. L. S. & Devay, J. A. (1971). The occurrence and characterization of a common antigen relationship between *Ustilago maydis* and *Zea mays*. *Physiol. Pl. Path.* **1**, 523–535.

Winoto Suatmadji, R. (1969). Studies on the effect of *Tagetes* species on plant-parasitic nematodes. *Wageningen, Fonds Landbouw. Export. Bur. Publ.* **47**, 1–132. Veenman & Zonen, N. V., Wageningen.

Wong, E. (1970). Structural and biogenetic relationships of isoflavonoids. *Prog. Chem. org. natur. Prod.* **28**, 1–73.

Wood, R. K. S. (1967). *Physiological plant pathology*. Blackwell, Oxford & Edinburgh. 570 pp.

Yang, S. L. & Ellingboe, A. H. (1972). Cuticle layer as a determining factor for the formation of mature appressoria of *Erysiphe graminis* on wheat and barley. *Phytopathology* **62**, 708–714.

Yarwood, C. E. (1954). Mechanism of acquired immunity to a plant rust. *Proc. natn. Acad. Sci. USA* **40**, 374–377.

Yarwood, C. E. (1956). Cross protection with two rust fungi. *Phytopathology* **46**, 540–544.

Yoder, O. C. (1972). Host-specific toxins as determinants of successful colonization by fungi. In *Phytotoxins in plant diseases*, pp. 457–460 (ed. R. K. S. Wood, A. Ballio & A. Graniti). Academic Press, London & New York.

Zaki, A. I., Keen, N. T., Sims, J. J. & Erwin, D. C. (1972). Vergosin and hemigossypol, antifungal compounds produced in cotton plants inoculated with *Verticillium albo-atrum*. *Phytopathology* **62**, 1398–1401.

Zentmyer, G. A. (1961). Chemotaxis of zoospores for root exudates. *Science* **133**, 1595–1596.

Index